Handbooks for the Identification of British Insects
Vol. 4. Part 5b

Keys to adults of
the water beetles
of Britain and Ireland (Part 2)

(Coleoptera: Polyphaga: Hydrophiloidea – both aquatic and terrestrial species)

Garth N. Foster

The Aquatic Coleoptera Conservation Trust

David T. Bilton

Plymouth University
Marine Biology & Ecology Research Centre

Laurie E. Friday

School of Physical Sciences,
University of Cambridge

Published for the Royal Entomological Society
The Mansion House
Bonehill
Chiswell Green Lane
Chiswell Green
St Albans
AL2 3NS
www.royensoc.co.uk

By the

Field Studies Council
Unit C1
Stafford Park 15
Telford
TF3 3BB
www.field-studies-council.org

ISBN: 978 0 90154 697 5

Contents

Dedication

This book is dedicated to the memory of Michael Hansen.

Abstract

Illustrated keys are provided for the families, genera, species and other taxa of British and Irish members of the superfamily Hydrophiloidea (Coleoptera suborder Polyphaga). The Hydrophiloidea comprise the families Helophoridae, Georissidae, Hydrochidae, Spercheidae, and the Hydrophilidae. Most of these beetles are aquatic and this book is the second part of the Handbook series concerning water beetles. However, this work includes terrestrial *Helophorus* and the many terrestrial species of the subfamily Sphaeridiinae of the Hydrophilidae. Notes include characters for distinguishing the sexes, basic information on biology and collecting methods, and reviews of distributions in Ireland and in Britain, including the Channel Isles. A key to all families of aquatic beetles is appended.

Acknowledgements

As ever the staff of many museums have proved a great help in allowing access to their collections, in particular: Roger Booth, Max Barclay and Christine Taylor of the Natural History Museum, London (NHM), where Robert Angus proved supportive again; Geoff Hancock of the Hunterian Museum of Zoology in the University of Glasgow, Dmitri Logunov of Manchester Museum in the University of Manchester, Richard Lyszkowski and Graham Rotheray of the Royal Scottish Museum, Granton. Magnus Sinclair loaned us some of his beautifully mounted beetles for photography.

Ron Carr checked material in Ken Side's collection in Maidstone Museum on our behalf. Don Elio Gentili clarified the current position of *Laccobius simulatrix*. Martin Hammond provided some interesting archival material. We are also grateful to Peter Hammond, Mark Telfer and Colin Welch for advice concerning the separation of two species of *Megasternum*. Oscar Vorst kindly provided guidance on *Cercyon castaneipennis*. Distribution data have been obtained from the many coleopterists and limnologists who have contributed to the recording schemes in Britain and Ireland.

This handbook has benefited from critical assessment by many people, in particular Martin Luff, for reading through an entire late draft, and participants on training courses – also the following for more specific recent assessments: John Bratton and Magnus Sinclair for *Cercyon*; Ron Carr for *Hydrochus*, *Laccobius* and *Sphaeridium*, Arno van Berge Henegouwen, for access to his unpublished key to *Sphaeridium*, also Jim Thomas for the same genus.

Rebecca Farley-Brown, Field Studies Council, is warmly thanked for her efforts in bringing this work to publication.

General introduction

Part 1 of *Keys to adults of the water beetles of Britain and Ireland* (Foster & Friday, 2011) concerned the "Hydradephaga", the aquatic members of one of the three suborders of the Coleoptera, the Adephaga. The Polyphaga is the largest suborder of Coleoptera and includes many families of beetles specialising in exploiting wetland habitats, mainly in the superfamilies Hydrophiloidea, Byrrhoidea and Scirtoidea. Part 2 within this Handbook series covers the Hydrophiloidea, the rest being planned to be covered in Part 3. The third suborder, the Myxophaga, is represented by a single diminutive species also to be covered in Part 3. Notes on amendments to Parts 1 and 2 will be made available in the journal *Latissimus* and at the website www.latissimus.org.

The old grouping, the "Palpicornia", comprised the superfamily Hydrophiloidea and the family Hydraenidae in the superfamily Staphylinoidea. The palpicorns provide an example of convergent evolution, their lengthened maxillary palps replacing the feeler function of the antennae, which are relatively short with a water-repellent club used to penetrate the water surface to permit renewal of the air supply. The apparent similarities between the two palpicorn groups are recognised here in a combined key to genera and to readily recognised individual species. Despite general agreement that the Hydrophiloidea and Hydraenidae evolved independently it is surprisingly difficult to find accessible characters to distinguish the hydraenid *Limnebius* from Hydrophilidae and the other hydraenid genera (*Hydraena*, *Enicocerus*, *Aulacochthebius*, and *Ochthebius*) from Hydrochidae. Their scientific separation is based on an array of rather obscure features such as are provided by the hind wings and the larvae, backed up by genetic data. More pragmatically, size matters here just as with the Hydradephaga, the Hydraenidae being largely recognised by their very small size.

It is hoped that this work will be found useful not only by water beetle specialists but also by limnologists in general. A key to the families of all wetland beetles is included (Key 15, p. 104) in an attempt to reduce the problems that someone, not necessarily an entomologist, might experience when dealing with an array of Coleoptera for the first time. Otherwise the format of this work follows the rest of the Handbook series. Streamlining for movement in water often effaces potential structural differences whereas colour and body size can be easily described and seen. Thus the keys do not necessarily follow systematic differences, with some subgroups within keys isolated by use of seemingly trivial characters rather than by orthodox morphology and systematically important characters. The descriptions applied here to genera may sometimes apply only to the one or two species by which the genus is represented in Britain and Ireland.

Beetles have complete metamorphosis with the egg developing into a larva that usually moults its cuticle three times (hence three instars) before pupating. Once emerged from the pupa and hardened the adult is fixed in size. Size thus provides a crucial identification character. If a beetle falls outside the range given for the species identified then it is likely that the identification is wrong. Measurements are from the tip of the elytra to the front of the head, taking into account the variations likely when the head is brought forward as in card-mounted specimens, or tilted downwards as in life.

Distributions of many species are given in some detail. This can build confidence in an identification, or sometimes lead to its destruction! If a species appears to have been found well outside the range recorded here then the identification should be double-checked and preferably confirmed by a specialist, and the record, if valid, publicised. Vice-county distributions are given only for the rarer species. Records from the offshore islands are given in the hope that these too will generate new information. Included are the Faroes, obviously not British but zoogeographically part of the British archipelago, and the Channel Isles, British of course politically but not part of the archipelago. The Channel Isles are of considerable interest in their own right.

Most of the drawings in this work are new. They were obtained by reworking with Adobe Photoshop the scanned traces of photographs taken with a digital camera attached to a Leica S8APO microscope. As such the relative dimensions of body parts should be real, with only a small amount of foreshortening associated with curved surfaces. This drawing process has emphasised how many characters offered for identification in the past *simply do not work*. With three exceptions characters that cannot easily be portrayed or measured have been rejected. The exceptions are colour, concave surfaces and a few optical illusions. Concave surfaces can barely be portrayed in photographs or drawings, but are readily apparent under a stereoscopic microscope. Some beetles show what appear to be structural differences, for example the relative widths of the head and the pronotum that can be based on differences in measurement so minuscule as to be useless. The human eye is often better than the ruler! If such differences can be drawn to retain the illusion then the character is available. Lastly, it is rarely possible to get a photograph in which the beetle is perfectly symmetrical. This can of course be because the beetle was not positioned properly, but often because the animal was slightly asymmetrical in the first place.

Photographs of beetles for the plates were taken with a digital camera fitted to a Leica Z6 APO macroscope. Image stacks were assembled using Helicon Focus software, and resulting composite images manually edited using Adobe Photoshop CS6.

A few of the choicer common names are included in the individual species accounts for those who might wish to use them. It should, however, be understood that their use in the absence of a Latin name will cause confusion. They are not in common parlance, as is the case for example with Odonata and the larger Lepidoptera. The Law of Priority does not apply to common names, a preferred name coming about by usage. A few of those listed here are recently conceived, but most are drawn from older publications.

A new classification for the Hydrophilidae

Short and Fikáček (2013) have developed a novel classification of the Hydrophilidae based on analyses of mitochondrial and nuclear genes, backed up by morphology. This became available after the draft of this handbook was in press. The new classification does not create any changes at the generic level in the British and Irish faunas, but it does lead to significant rearrangements at tribe and subfamily levels as summarised below:

    Subfamily Hydrophilinae Latreille, 1802

        Tribe Berosini Mulsant, 1844 – *Berosus*

        Tribe Laccobiini Bertrand, 1954 – *Paracymus, Laccobius*

        Tribe Hydrophilini Latreille, 1802 – *Hydrophilus, Hydrochara*

        Tribe Hydrobiusini Mulsant, 1844 – *Limnoxenus, Hydrobius*

    Subfamily Chaetarthriinae Bedel, 1881

        Tribe Chaetarthriini Bedel, 1881 – *Chaetarthria*

        Tribe Anacaenini Hansen, 1991 – *Anacaena*

    Subfamily Enochrinae Short & Fikáček, 2013 – *Cymbiodyta, Enochrus*

    Subfamily Acidocerinae Zaitzev, 1908 – *Helochares*

    Subfamily Sphaeridiinae Latreille, 1802

        Tribe Coelostomatini Heyden, 1891 – *Coelostoma, Dactylosternum*

        Tribe Sphaeridiini Latreille, 1802 – *Sphaeridium*

        Tribe Megasternini Mulsant, 1844 – *Megasternum, Cryptopleurum, Cercyon*

Checklist

Classification here follows Michael Hansen's (1991, 1999) treatment of the Hydrophiloidea, the names of species being updated following Löbl & Šmetana (2004, 2011). The subgenera of *Helophorus* follow Angus (1992a) but where necessary retaining Kuwert's (1886) original spellings.

Suborder Polyphaga Emery, 1886

Superfamily Hydrophiloidea Latreille, 1802

Family HELOPHORIDAE Leach, 1815

HELOPHORUS Fabricius, 1775
    Subgenus EMPLEURUS Hope, 1838
        nubilus Fabricius, 1777
        porculus Bedel, 1881
        rufipes (Bosc d'Antic, 1791)
    Subgenus KYPHOHELOPHORUS Kuwert, 1886
        tuberculatus Gyllenhal, 1808
    Subgenus TRICHELOPHORUS Kuwert, 1886
        alternans Gené, 1836
    Subgenus HELOPHORUS Fabricius, 1775
        aequalis Thomson, 1868
            aquaticus by usage, not Linnaeus, 1758
        grandis Illiger, 1798
    Subgenus ATRACTOHELOPHORUS Kuwert, 1886
        arvernicus Mulsant, 1846
        brevipalpis Bedel, 1881
            guttulus not Motschulsky, 1860
    Subgenus RHOPALOHELOPHORUS Kuwert, 1886
        dorsalis (Marsham, 1802)
        flavipes Fabricius, 1792
        fulgidicollis Motschulsky, 1860
        granularis (Linnaeus, 1760)
        griseus Herbst, 1793
        laticollis Thomson, 1853
        longitarsis Wollaston, 1864
        minutus Fabricius, 1775
        nanus Sturm, 1836
        obscurus Mulsant, 1844
        strigifrons Thomson, 1868

Family GEORISSIDAE Castelnau, 1840

GEORISSUS Latreille, 1809
        crenulatus (Rossi, 1794)

Family HYDROCHIDAE Thomson, 1859

    HYDROCHUS Leach, 1817
        angustatus Germar, 1824
        brevis (Herbst, 1793)
        crenatus (Fabricius, 1792)
            carinatus Germar, 1824
        elongatus (Schaller, 1783)
        ignicollis Motschulsky, 1860
        megaphallus van Berge Henegouwen, 1988
        nitidicollis Mulsant, 1844

Family SPERCHEIDAE Erichson, 1837

    SPERCHEUS Kugelann, 1798
        emarginatus (Schaller, 1783)

Family HYDROPHILIDAE Latreille, 1802

    Subfamily HYDROPHILINAE Latreille, 1802

    ANACAENA Thomson, 1859
        bipustulata (Marsham, 1802)
        globulus (Paykull, 1829)
        limbata (Fabricius, 1792)
        lutescens (Stephens, 1829)

    PARACYMUS Thomson, 1867
        aeneus (Germar, 1824)
        scutellaris (Rosenhauer, 1856)

    BEROSUS Leach, 1817
        Subgenus BEROSUS Leach, 1817
        affinis Brullé, 1835
        luridus (Linnaeus, 1760)
        signaticollis (Charpentier, 1825)
        Subgenus ENOPLURUS Hope, 1838
        fulvus Kuwert, 1888
        not *spinosus* (von Steven, 1808)

    CHAETARTHRIA Stephens, 1835
        seminulum (Herbst, 1797)
        simillima Vorst & Cuppen, 2003

    CYMBIODYTA Bedel, 1881
        marginellus (Fab.)
        marginella unjustified emendation

ENOCHRUS Thomson, 1859
  Subgenus ENOCHRUS Thomson, 1859
    melanocephalus (Olivier, 1793)
  Subgenus LUMETUS Zaitzev, 1908
    bicolor (Fabricius, 1792)
    fuscipennis (Thomson, 1884)
    halophilus (Bedel, 1878)
    ochropterus (Marsham, 1802)
    quadripunctatus (Herbst, 1797)
    testaceus (Fabricius, 1801)
  Subgenus METHYDRUS Rey, 1885
    affinis (Thunberg, 1794)
    coarctatus (Gredler, 1863)
    nigritus Sharp, 1872
      isotae Hebauer, 1981

HELOCHARES Mulsant, 1844
    lividus (Forster, 1771)
    obscurus (Müller, 1776)
    punctatus Sharp, 1869
      griseus by usage, not (Fabricius, 1787)
      obscurus by usage, not (Müller, 1776)

HYDROBIUS Leach, 1815
    fuscipes (Linnaeus, 1758)
      chalconatus Stephens, 1829
      aeneus Solier, 1834
     var. *subrotundus* Stephens, 1829
      picicrus Thomson, 1865
      subrotundatus auctt.
     var. *rottenbergi* Gerhardt, 1872

LIMNOXENUS Motschulsky, 1853
    niger (Gmelin, 1790)

HYDROCHARA Berthold, 1827
    caraboides (Linnaeus, 1758)

HYDROPHILUS Geoffroy, 1762
    piceus (Linnaeus, 1758)

LACCOBIUS Erichson, 1837
  Subgenus DIMORPHOLACCOBIUS Zaitzev, 1938
    atratus Rottenberg, 1874
    bipunctatus (Fabricius, 1775)
    simulatrix d'Orchymont, 1936
      simulator d'Orchymont unjustified emendation
    sinuatus Motschulsky, 1849
    striatulus (Fabricius, 1801)
     var. *purpurascens* Newbery, 1908

    ytenensis Sharp, 1910
      atrocephalus in part Reitter, 1872
   Subgenus LACCOBIUS Erichson, 1837
    colon (Stephens, 1829)
      biguttatus Gerhardt, 1877
    minutus (Linnaeus, 1758)

Subfamily SPHAERIDIINAE Latreille, 1802

COELOSTOMA Brullé, 1835
    orbiculare (Fabricius, 1775)

DACTYLOSTERNUM Wollaston, 1854
    abdominale (Fabricius, 1792)

CERCYON Leach, 1817
   Subgenus CERCYON Leach, 1817
    alpinus Vogt, 1969
    bifenestratus Küster, 1851
    convexiusculus Stephens, 1829
      intermixtus Sharp, 1918
    depressus Stephens, 1829
    granarius Erichson, 1837
    haemorrhoidalis (Fabricius, 1775)
      aquatilis Donisthorpe, 1932
    impressus (Sturm, 1807)
      atomarius (Fabricius, 1775)
    lateralis (Marsham, 1802)
    littoralis (Gyllenhal, 1808)
      litorale Thomson, 1860 not Gyllenhal
     var. *binotatus* Stephens, 1829
     var. *ruficollis* Schilsky, 1888
    marinus Thomson, 1853
    melanocephalus (Linnaeus, 1758)
    nigriceps (Marsham, 1802)
      atricapillus (Marsham, 1802)
    obsoletus (Gyllenhal, 1808)
      lugubris (Olivier, 1790), not (Fourcroy, 1785)
    pygmaeus (Illiger, 1801)
    quisquilius (Linnaeus, 1760)
    sternalis (Sharp, 1918)
    terminatus (Marsham, 1802)
    tristis (Illiger, 1801)
    unipunctatus (Linnaeus, 1758)
     var. *impunctatus* Kuwert, 1890
     var. *janssoni* Nyholm, 1952
   Subgenus DICYRTOCERCYON Ganglbauer, 1904
    ustulatus (Preyssler, 1790)
   Subgenus PARACERCYON Seidlitz, 1888
    analis (Paykull, 1798)

Subgenus PARACYCREON d'Orchymont, 1942
laminatus Sharp, 1873
MEGASTERNUM Mulsant, 1844
concinnum (Marsham, 1802)
boletophagum by usage, not (Marsham, 1802)
obscurum (Marsham, 1802)
immaculatum (Stephens, 1829)

CRYPTOPLEURUM Mulsant, 1844
crenatum (Kugelann, 1794)
minutum (Fabricius, 1775)
subtile Sharp, 1884

SPHAERIDIUM Fabricius, 1775
bipustulatum Fabricius, 1781
lunatum Fabricius, 1792
marginatum Fabricius, 1787
scarabaeoides (Linnaeus, 1758)

Superfamily Hydrophiloidea Latreille

Introduction to the Hydrophiloidea

One of the few almost constant characters of the Hydrophiloidea is that the maxillary palps are held further forward than the antennae, which are typically tucked beneath the head. In many hydrophiloids the palps are also as long as or longer than the antennae, although the opposite may also be true as in *Georissus*. Hydrophiloid antennae generally bear a 3-segmented hairy club on a 4-8 segmented stem, an exception being *Spercheus*, which has two more of the segments hairy in a complex structure. In addition to the space beneath the elytra aquatic hydrophiloids can carry a large air bubble on the underside, which is renewed using the water-repellent antennal clubs. Thus hydrophiloid beetles will surface head first whereas the Hydradephaga surface rear end first. Swimming ability varies, some being as active and effective as the Hydradephaga, but a few can barely dive and the majority cannot swim at all, walking just below the surface, sometimes upside down making the silvery bubble prominent.

There is controversy concerning the internal classification of this the main superfamily of the aquatic Polyphaga just as there is with the overall classification of the suborder. Here the Hydrophiloidea is regarded as having six families following Hansen (1991, 1999) as endorsed by Short and Fikáček (2011). Five of these families occur in Britain and Ireland: Helophoridae, Georissidae, Hydrochidae, Spercheidae and Hydrophilidae sensu stricto. Beutel and Leschen (2005) and Bouchard *et al.* (2011) reduce these to subfamilies of an enlarged Hydrophilidae.

Several members of the Helophoridae and many of the Sphaeridiinae of the Hydrophilidae are not aquatic. These allies of water beetles are included in this volume to retain systematic integrity in the Handbook series. The wholly terrestrial Sphaeritidae Shuckard and Histeridae Gyllenhal are sometimes included in the Hydrophiloidea, although they are now most commonly referred to a separate superfamily, the Histeroidea (see Halstead, 1963).

Adults of the aquatic Hydrophiloidea are prominent members of the freshwater fauna, but their larvae are rarely seen, many being cryptic and semi-terrestrial. This is in contrast to the Hydradephaga where both adults and larvae are fully aquatic with larvae of many species found commonly during the breeding season.

Our understanding of the Hydrophiloidea has changed greatly since the publication of Balfour-Browne (1958). For example Hodge and Jones (1995) noted fourteen species brought forward comparatively recently, plus the complete overhaul of *Helophorus* in a series of publications culminating in Angus (1992a). It is likely that species complexes will continue to be exposed in some hydrophilid genera by genetic analysis and by reappraisal of morphology, as has been hinted here in treatments of *Hydrobius*, *Enochrus*, *Anacaena*, *Paracymus*, and *Megasternum*.

Morphology and field characters

Some of the aquatic beetles in the Hydrophiloidea that swim have a smooth outline rather like Diving Beetles (Dytiscidae, suborder Adephaga), but the metacoxae are never modified and fused with the underside as in the Adephaga. Unlike the Hydradephaga the Hydrophiloidea come to the surface head first to renew the air supply using their antennae. Most hydrophiloids do little more than move just below the water surface dorsal side uppermost or, hanging from the surface film, walk rapidly, ventral side uppermost with their air bubble prominent. Genera dominated by non-swimming species such as *Helophorus* and *Laccobius* include a few species that can dive, providing a useful field character. *Berosus*, *Hydrochara*, and *Hydrophilus* are all active divers. The provision of long swimming hairs on the legs correlates with swimming ability. The term "tarsal formula" refers to the number of segments on each tarsus from front to rear. Most Hydrophiloidea are typically 5:5:5, though *Cymbiodyta* is 5:4:4, and males of *Berosus* are 4:5:5.

The expressions "fore", "mid" and "hind" are used interchangeably with both first, second and third and "pro-", "meso-" and "meta-", the latter suffixes often being more convenient to use than the ordinary words. Structures with technical names are labelled on figures.

With some reluctance the term "stria", which should really refer to a groove alone, has been used for the entirety of each row on the elytra, whether grooved or marked only by a row of punctures (Fig. 1). The extent to which each "stria" is truly grooved is important in characterising several hydrophilid genera. *Hydrobius*, *Berosus* and some *Helophorus* have a short, extra "intercalary stria" next to the "sutural stria", i.e. that next to the elytral suture. If the elytra are gaping it is also possible to see the grooves and ridges that together normally lock the elytra together, and these structures can sometimes be confused with elytral striae.

The space between the "striae", again important for identification, is often described as an interstrial space, but the term "interstice" is more neutral and can be applied with confidence when there are no real striae at all!

Most punctures bear some kind of hair or bristle (generally, a "seta") but few long hairs can be seen with the ordinary binocular microscope. The distribution of the "setiferous punctures" that bear these hairs is often independent of the main rows of punctures, and can also be important for identification. Most elytra also have much finer punctures and other ornamentation to be discussed under the respective genera.

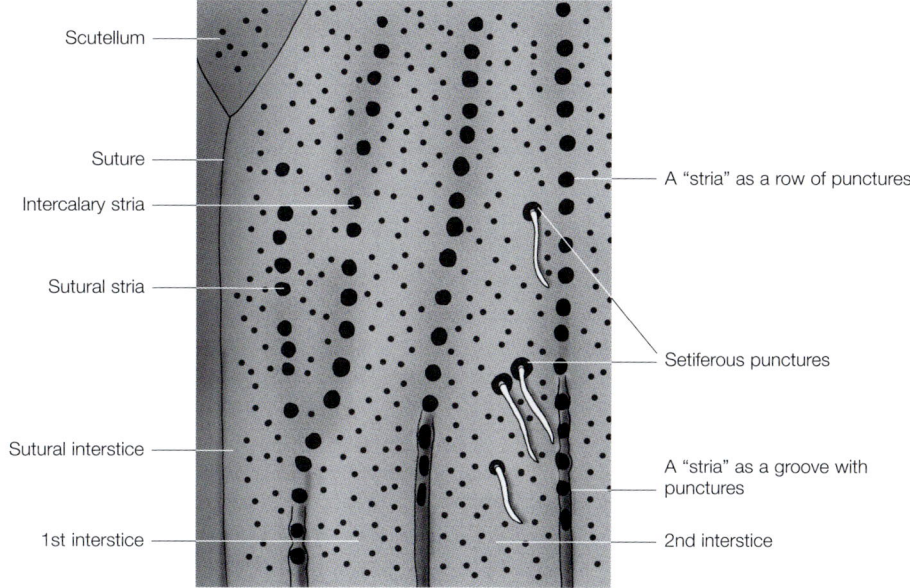

Scutellum

Suture

Intercalary stria

Sutural stria

A "stria" as a row of punctures

Setiferous punctures

Sutural interstice

A "stria" as a groove with punctures

1st interstice

2nd interstice

Figure 1. Features of the front of a hydrophilid elytron.

Hydrophiloid genitalia

Examination of the male genitalia is essential to confirm identification in many cases. External sexual characters are often weak or absent, but include enlarged tarsal claws and/or wider tarsal segments in males of most genera, such features being useful in deciding which specimens to dissect. Also, females might be recognised by the extremities of the genitalia, two processes with hairy tips, revealing themselves without the need for a full dissection. The genitalia are most easily removed by gently squeezing the freshly killed animal. The aedeagophore is symmetrical with three processes fitted onto a basal piece. It can usually be extracted without causing damage by gripping it with fine forceps. It should be mounted in a drop of fluid such as DMHF (dimethyl hydantoin formaldehyde) as it is important to prevent drying out, when it might become irrevocably distorted.

The main part of the external genitalia of males, the aedeagophore, frequently provides good characters for identification, its structure being best described with reference to particular genera. The generalised structure is trilobed and quadripartite, a median lobe being flanked by a pair of parameres and mounted on a basal piece. In the species covered in Part 2 it is mostly flat and transparent enough for structures to be seen regardless of whether the aedeagophore is viewed from the dorsal or ventral side. The naming of supporting struts and other detail varies with the genus and its expert's view, but some uniformity of naming such parts has been attempted here.

Collecting methods for Hydrophiloidea

The traditional equipment of choice in Britain and Ireland to catch water beetles is the pond net, preferably with a D-framed or circular rim strong enough to push aside vegetation, and with the fabric edge of the net bag protected inside the main rim: the mesh of the bag is typically 1 mm. However, kitchen sieves are popular in other countries and can be just as effective for the smaller beetles. Almost as important as the net or sieve is the sorting tray, which should be as large as fieldwork constraints will allow. The first sweep through the water is often the only one likely to capture the larger beetles, which are more amenable to capture by trapping. However, the majority of the Hydrophiloidea are often best caught by stirring up sediment and vegetation in shallow water, or by splashing the bank or dunking marginal vegetation and then scooping up the beetles that have floated to the surface.

Hydrophiloidea will also be intercepted in light traps and pitfall traps, also, on shiny surfaces such as car roofs. A problem here is that many individuals undertaking flights will be teneral, lacking fully developed pigmentation and with the genitalia weakly sclerotised and easily distorted. Underwater traps, such as are used for Dytiscidae, are more limited in value for the less active hydrophilids.

Species living in litter and dung must be sought by sifting, shaking or otherwise disturbing the material, preferably over a plastic tray. Some Sphaeridiinae, particularly *Sphaeridium* species, live in semi-liquid dung such as fresh cow pats: they can be sought using a spoon or by stirring the material in water.

Key 1. Hydrophiloidea and Hydraenidae

A key that covers all beetle families with aquatic representatives is provided in Appendix 1 (Key 15, p. 104), along with guidance on water beetles for the non-specialist.

1. Length 38-48 mm; colour silvery black; underside with a keel in the midline extending to the second of the five visible abdominal sternites, and ending in a long, sharp point (Fig. 2) ....
.... Hydrophilidae, Hydrophilinae
11. ***Hydrophilus*** Geoffroy
1. ***Hydrophilus piceus*** (Linnaeus) (p. 66)

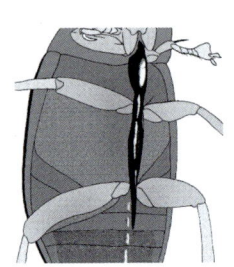

2

- Less than 20 mm long; variously coloured; if with a keel then this does not extend beyond the hind coxae 2

2. Length 15-18 mm; black; underside with a keel in the midline, present on the prosternum and running from the mesosternum to be level with the hind coxae (Fig. 3) ... Hydrophilidae, Hydrophilinae
10. ***Hydrochara*** Berthold
1. ***Hydrochara caraboides*** (Linnaeus) (p. 65)

3

- Length 11 mm or less; variously coloured; keel if present not as above 3

3. Pronotum with five grooves running from front to rear (Fig. 4) .....
.... Helophoridae
1. ***Helophorus*** Fabricius (20 spp.) (p. 18)

A few terrestrial species of *Helophorus* in the subgenus *Empleurus* (p. 28) may have the outer grooves not clearly separated from the outer intervals.

- Pronotum without such grooves 4

intercalary stria   Scutellum

4

4. Head hidden beneath a ridged forward extension (or "shelf") of the pronotum (Fig. 5); scutellum not visible when the elytra are closed; length 1.4-2.1 mm .... Georissidae

1. *Georissus* Latreille

1. *Georissus crenulatus* (Rossi) (p. 34)

A semi-terrestrial and cryptic species, often covered with mud.

- Pronotum not as in Fig. 4; scutellum visible as in Figs 4 and 6 ... 5

5

5. Pronotum narrowing in its rear half, body usually covered with dents or grooves or coarse irregular punctures, often flat or elongate 6

- Pronotum not narrowing in its hind half; body usually smooth and domed and rounded 8

6. Length 5.5-7.0 mm; front of the head with a deep, wide indentation (Fig. 6); surface of the pronotum uneven, sometimes with a large shallow depression on either side positioned as indicated in Fig. 6 (➘); elytra yellowish with brown blotches .... Spercheidae

1. *Spercheus* Kugelann

1. *Spercheus emarginatus* (Schaller) (p. 40)

A species known only from England and probably now extinct.

- Length 1.0-4.5 mm; front of head smoothly rounded, or with a narrow V-shaped notch, or a very shallow indentation; pronotum either smooth, or with depressions accentuated by their surrounding ridges, or with narrow grooves; elytra generally dark 7

6

7. Length 2.2-4.5 mm; pronotum with five large depressions, three in front, two behind, rendered most noticeable by their smooth rims (Fig. 7), the pronotal surface otherwise being covered with pits on a roughened surface .... Hydrochidae

1. *Hydrochus* Leach (7 spp.) (p. 35)

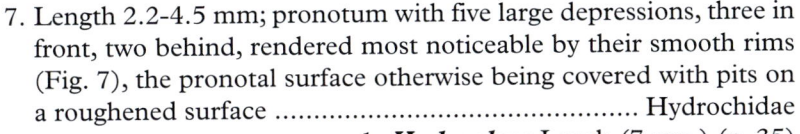

7

- Length 1.0-2.8 mm; pronotum with many different arrangements of grooves and pits but never as above, in particular lacking the larger central depression, though often with a central groove (Figs 9-11) .... Hydraenidae

Hydraena, Aulacochthebius,
Enicocerus and *Ochthebius*
Part 3

Examples of the grooving of the pronotum are here provided by *Hydraena gracilis* Germar (Fig. 8), *Aulacochthebius exaratus* (Mulsant) (Fig. 9), *Enicocerus exsculptus* (Germar) (Fig. 10), and *Ochthebius minimus* (Fabricius) (Fig. 11).

 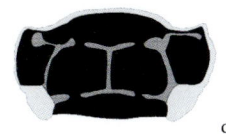  

8   9   10   11

8. Elytra truncate with protruding rear of abdomen bearing a pair of tufts of long hairs in both sexes (Fig. 12); length 1.0-2.5 mm .... Hydraenidae

Limnebius

Part 3

Part of the abdomen may protrude in hydrophilid beetles distended in weak preservative, but the elytra are always rounded towards the tip, not truncated.

- Elytra rounded towards the tip, usually without the abdomen visible; length 1 mm upwards 9

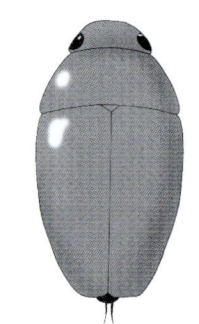

12

9. Maxillary palps shorter than the antennae, which have a compact club (see Plates 69-72, 74-90, 92-102); second segment of the maxillary palps dilated (Fig. 13, maxillary palp and antenna of *Cercyon ustulatus* viewed from below); first (basal) segments of middle and hind tarsi longer than their second segments (Fig. 14, which shows the metatarsus of *C. ustulatus* from above); elytra, if striate, then without intercalary striae .... Hydrophiloidea, Hydrophilidae, Sphaeridiinae

10

Mostly terrestrial species in dung and rotting plant material.

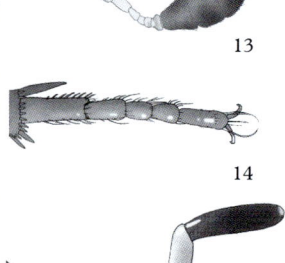

13

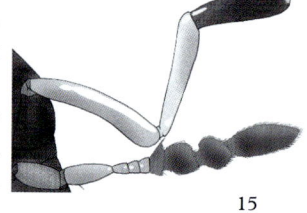

14

- Maxillary palps as long as or longer than the antennae, which have looser clubs (Fig. 15, underside of the head of *Enochrus nigritus*); second segment of the maxillary palps rarely dilated; either first segments of middle and hind tarsi shorter than the second (Fig. 16, as in *Paracymus aeneus* – sometimes hidden by the spines at the tip of the tibiae, as in *Helochares lividus* (Fig. 17), or tarsus with only four segments (*Cymbiodyta*, Fig. 18); elytra often striate and with or without intercalary striae (Fig. 33) .... Hydrophiloidea, Hydrophilidae, Hydrophilinae

15

15

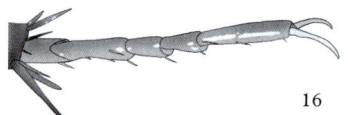

16

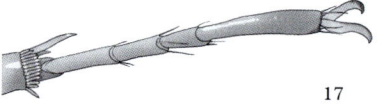

17

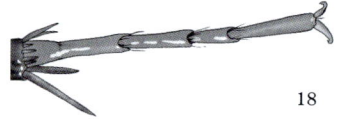

18

Mainly aquatic species.

An intercalary stria is here defined as a short row of punctures, typically in the space (the first interstice) between the sutural (or first) main stria and the second one (see Figs 4 and 33). Sometimes this short row of punctures occurs between the first stria and the suture (see Fig. 1), when it might be called the scutellary stria, a term not used in this work.

10. Scutellum decidedly longer than wide (Fig. 19); large species, 4.2-7.7 mm long; black, usually with red and yellow markings on the elytra (Plates 99-102); antennae 8-segmented (Fig. 20) .... 1. **Sphaeridium** Fabricius (4 spp.)(p. 94)

- Scutellum about as long as wide; smaller species, 1.4-5.0 mm; various colour patterns, not usually spotted; antennae usually with 9 segments, the extra one being amongst the small segments in the middle 11

19

20

11. Side of head with a notch in front of the eye, exposing the base of the antenna (Fig. 21), but the eye itself not notched (Fig. 22); variously coloured; length 1.4-4.2 mm 12

- Side of head without a notch, the base of the antenna being concealed (Fig. 23); front edges of eyes notched ("emarginate" - Fig. 24); black; length 3.8-5.0 mm 14

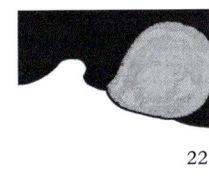
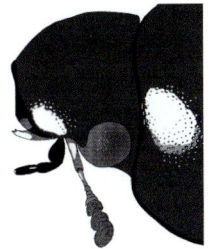
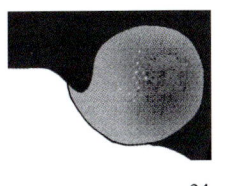

21   22   23   24

12. Prosternum protruding ventrally as a ridge (Fig. 25); elytral epipleurs broadly visible, at least in the front third (as in *Cercyon lateralis*, Fig. 26, also illustrated in black in the inset) .... 15. **Cercyon** Leach (22 spp.) (p. 73)

- Prosternum developed as a broad plate (Fig. 27); elytral epipleurs narrow except at base (as in *Cryptopleurum subtile*, Fig. 28) 13

"Epipleur" as used here refers to the whole of that part of the elytron folded to face ventrally. Strictly speaking this is made up of the inner true epipleuron and an additional strip outside it known as the pseudepipleuron. Also potentially confusing is the extent to which the sides of the elytra are visible from below, broadly in Fig. 28 but scarcely at all in Fig. 26.

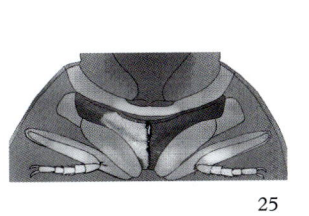
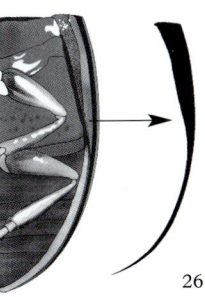
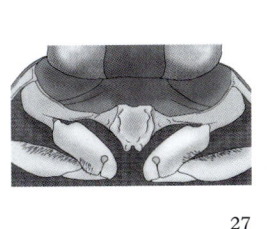
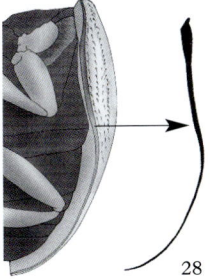

25   26   27   28

13. Dorsal surface without obvious hair; outer face of fore tibia with a distinct excision (Fig. 29) .... 16. **Megasternum** (2 taxa) (p. 90)

- Dorsal surface with hairs; outer face of fore tibiae without excision (Fig. 30) .... 17. **Cryptopleurum** (3 spp.) (p. 92)

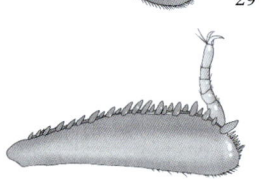

14. Elytra evenly covered with punctures, the only grooves being those either side of the suture (the sutural striae) running forward for about half of the elytral length (Fig. 31); body shape strongly rounded (Fig. 31); length 4.0-4.8 mm .... 13. **Coelostoma** Brullé
  1. Coelostoma orbiculare (Fabricius) (p. 72)

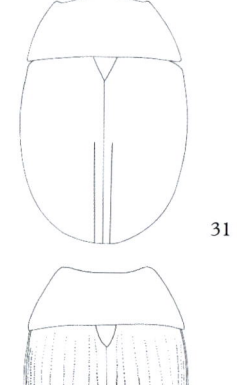

- Elytra with 11 punctured striae, the innermost next to the suture being grooved from the rear to in front of the middle (Fig. 32); body shape slightly more elongate (Fig. 32); length 3.8-5.0 mm .... 14. **Dactylosternum** Wollaston
  1. Dactylosternum abdominale (Fabricius) (p. 73)

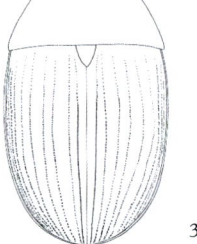

15. Scutellum distinctly longer than wide (↗ Fig. 33); middle and hind legs with long swimming hairs on both the tibiae and the tarsi (Fig. 34); mesosternum with a dark hairless pit in the midline between hairy prominences (Fig. 35); elytra with longitudinal punctured striae running from near the front to the rear, plus the intercalary stria, an additional and short row of punctures between the 1st and 2nd rows (Fig. 33, see also footnote on p. 12), and other punctures scattered between striae; upper side brown, grey or, in life, sometimes green; appearance characteristic (Plates 36-39); fast-swimming insects often stridulating audibly; length 3.5-5.5 mm ...... Hydrophilidae, Hydrophilinae
  3. **Berosus** (4 spp.) (p. 46)

- Scutellum about as wide as long; swimming hairs, if visible, confined to middle and hind tarsi; mesosternum without a central pit; elytra with or without intercalary striae; colour usually brown or black, sometimes an olive green; more likely to dog-paddle upside down just below the surface if disturbed, but a few species can dive; not likely to stridulate audibly; length 1-11 mm 16

When dry the swimming hairs lay as a slick along the leg but they will fan out if wetted.

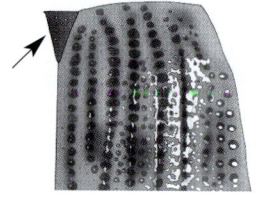

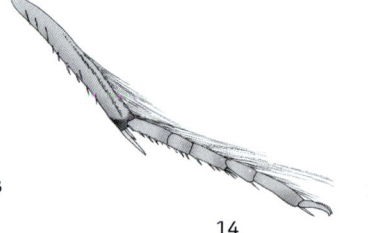

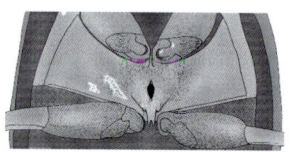

16. Body entirely black; length 6 mm or more 17

- Body of various colours; length less than 6 mm 18

17. Length 6.4-8.0 mm; elytra each with 10 punctured striae most strongly grooved at the rear, fading towards weak series of punctures at the front where intercalary series of punctures are also visible (Fig. 36); underside without keel-like processes on mesosternum and metasternum ....
.... Hydrophilidae, Hydrophilinae
1. *Hydrobius* Leach
1. *Hydrobius fuscipes* **(Linnaeus)** (p. 63)

- Size 8.0-9.8 mm; the elytral rows of punctures not in grooved striae, except for the rear half of the rows either side of the suture, which form distinct sutural striae (Fig. 37); mesosternum with a median keel, and front of metasternum also keeled (Fig. 38) ....
.... Hydrophilidae, Hydrophilinae
1. *Limnoxenus* Motschulsky
1. *Limnoxenus niger* **(Gmelin)** (p. 64)

18. Globular insects less than 2 mm long, often partly curled up (conglobated) (Fig. 39); first two visible abdominal sternites covered with a fringe of long hairs retaining blocks of gelatinous material (dislodged in Fig. 40); body entirely shining black ....
.... Hydrophilidae, Hydrophilinae
4. *Chaetarthria* (2 spp.) (p. 49)

The white areas in Fig. 39 represent highlights, not pale spots.

- With a strongly convex upper surface but flat underneath and incapable of rolling-up; abdominal sternites not fringed as above; body variously coloured, at least with margins paler 19

19. Elytra lacking any striae, even the sutural striae being absent throughout 20

- Each elytron with at least a sutural stria in the rear half 21

20. Maxillary palps about twice as long as the antennae (Fig. 41); larger insects (4.4-6.0 mm long); upper surface drab yellow, olive or brown without any metallic reflections; pronotum and elytra not contrasting strongly in colour; hind tibiae straight ....
.... Hydrophilidae, Hydrophilinae
7. *Helochares* Mulsant (3 spp.)(p. 62)

- Maxillary palps about the same length as the antennae (Fig. 42); smaller insects (2.4-4.2 mm long); upper surface mainly yellowish grey with head and a central patch on the pronotum black, the elytra flecked with dark marks and sometimes largely blackened and/or with metallic purple or green reflections; hind tibiae slightly curved inwards (Fig. 43), giving a bow-legged appearance .... Hydrophilidae, Hydrophilinae
12. *Laccobius* Erichson (8 spp.)(p. 66)

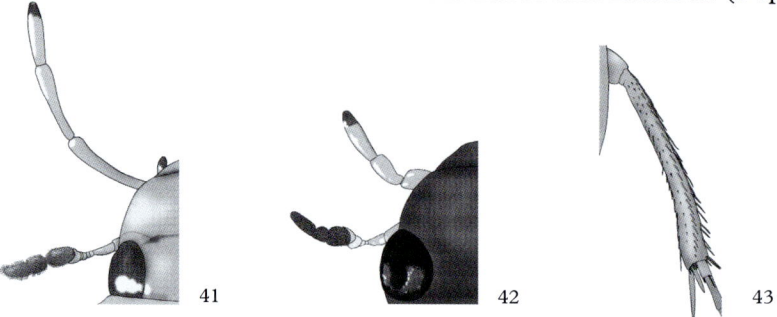

21. First visible (from above) segment of maxillary palp distinctive, being long, curved outwards on the inner side and concave or almost straight on the outer side (Fig. 44) ....
.... Hydrophilidae, Hydrophilinae
6. *Enochrus* Thomson (10 spp.)(p. 52)

- First visible segment of maxillary palp straight or slightly bent inwards (Fig. 45), or mostly hidden under the head (Fig. 46) 22

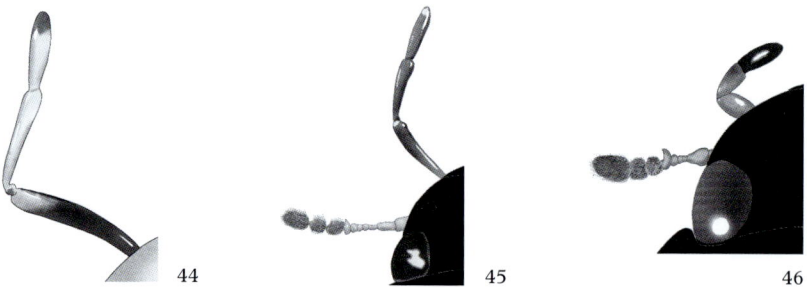

Enochrus melanocephalus is somewhat intermediate in the maxillary palp character (see Fig. 164, p. 53) and is otherwise easily distinguished by the strong contrast between the black head and the orange or yellow pronotum.

22. Maxillary palps as long as or shorter than the antennae (e.g. as in *Paracymus scutellaris*, Fig. 46); tarsal formula 5:5:5, i.e. normal for Hydrophilidae; small beetles up to 3.3 mm long 23

- Maxillary palps longer than antennae (Fig. 45); tarsal formula 5:4:4, i.e. mid and hind tarsi each with four segments unlike any other hydrophilid (Fig. 47); black upper side with broad yellowish brown margins to pronotum and elytra; length 3.3-4.3 mm .... Hydrophilidae, Hydrophilinae
5. *Cymbiodyta* Bedel
1. *Cymbiodyta marginellus* (Fabricius) (p. 51)

The basal segments of the mid and hind tarsi of *Cymbiodyta* can be seen (Fig. 47 ✗) to have formed from fusion of two segments.

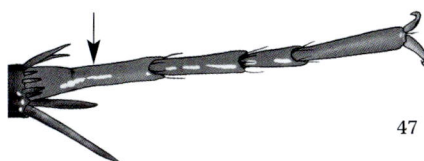

47

23. Upper surface uniformly dark with metallic reflections; hind femora entirely shining when viewed from below .... Hydrophilidae, Hydrophilinae
2. *Paracymus* Thomson (2 spp.)(p. 44)

- Upper side black or brown, sometimes spotted and with margins usually paler than the rest, but without metallic reflections; hind femora with hairy covering making at least their inner halves dull (Fig. 48) .... Hydrophilidae, Hydrophilinae
1. *Anacaena* Thomson (4 spp.)(p. 41)

48

Family HELOPHORIDAE Thomson

The Helophoridae may be dated back to the late Jurassic, their ancestors being truly aquatic as adults with terrestrial forms developed later (Fikáček *et al.*, 2012).

1. *HELOPHORUS* Fabricius

The pronotum, with its median groove and two grooves on each side (Fig. 49), will separate *Helophorus*, the single genus in the Helophoridae, from all other beetles. Body length varies from 1.9 to 7.5 mm but the body shape, elongate and never globular, sets the genus apart from other hydrophiloids except *Hydrochus*. The ornamentation of the pronotum is also characteristic, the intervals being covered with bun-shaped granules that emphasise the grooving, as they also do the Y-shaped groove on the head. Variations in metallic colouration – golden, bronze, green, red, purple, or blue – emphasise the structure of the head and pronotum. A metallic finish, usually greenish, extends onto the elytra of some species. The primary punctures of the elytra are large and organised into rows. The intervening spaces, or interstices, may have small punctures bearing bristles (as in the terrestrial species) or fine hairs. Every other interstice, sometimes all of them, may be raised with the main rows of punctures in grooves here known as striae. Patterns on the elytra may be useful for identification though they are quite variable within species. Most *Helophorus* have at the very least dark spots two-thirds of the way down the elytra together making an inverted V.

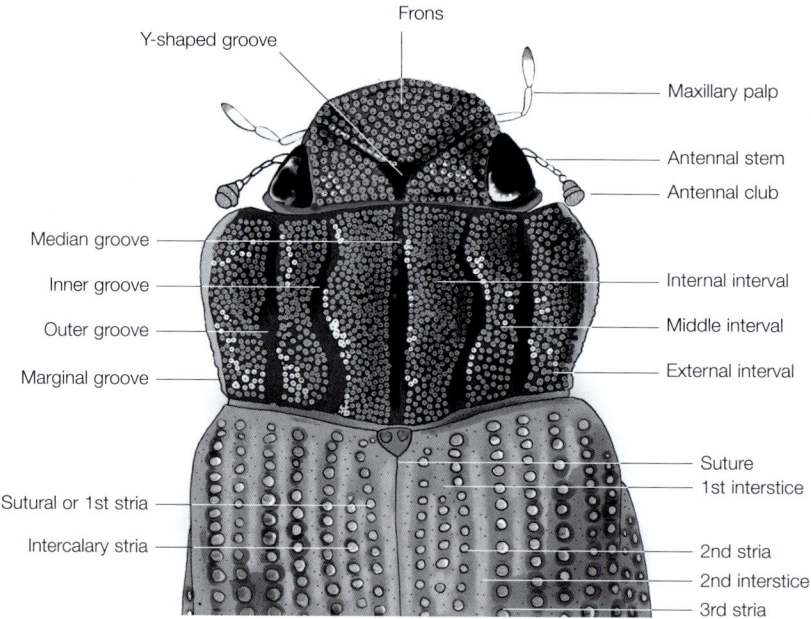

Figure 49. Key features of the upper side of an *Helophorus*.

Helophorus adults feed on decaying vegetable matter whereas most of their larvae are predatory, living at the water's edge and hauling prey out of the water. The three species in the subgenus *Empleurus* are exceptional in being largely terrestrial throughout their life with both adults and larvae phytophagous. All of the aquatic species can move by paddling with the dorsal side uppermost or by walking upside down just below the water's surface with their air bubbles prominent: a few species can swim more effectively, and the ability to dive may even provide a field character. Long, fine hairs will be found on the legs of the swimming species, but these can be obscured by glue if a specimen is card-mounted. Many *Helophorus* appear to retain the ability to fly throughout their adult life, some species being commonly intercepted on reflective surfaces such as car roofs; *H. granularis* is an exception, having a form with reduced wings.

The difference between symmetrical and asymmetrical maxillary palps must be mastered because the former, found in *H. brevipalpis*, will account for most captures of smaller species. Any series of identifications that fails to generate at least the occasional *H. brevipalpis* should be viewed with suspicion.

The sexes cannot be distinguished by external characters, though males are generally smaller than females, so much so in a few cases that separate species might be suspected. The two bristle-bearing tips of the female genitalia often protrude enough to allow them to be sexed without dissection. Otherwise, with freshly killed specimens, the beetle's abdomen should be pressed gently to extrude the male genitalia. The abdomens of dry preserved specimens should be softened in hot water to facilitate removal of the aedeagophore. It is sometimes impossible to separate with any certainty females of *H. flavipes* from those of *H. obscurus*, and *H. minutus* from those of *H. griseus* – fieldwork should take this into account, ensuring enough material is taken to capture males.

The structure of the aedeagophore (Fig. 50) can easily be seen when fixed to the mounting card in an aqueous mountant such as DMHF. The aedeagophore's parts are the basal piece and the median lobe, which is flanked by a pair of parameres. The median lobe can itself be divided into the tube and the struts, the relative lengths of the lobe and struts providing an important identification guide. The orifice of the genitalia lies just below the tip of median lobe and is supported by a strengthening ring (the annulus or corona) and a spur. The shape of the parameres, and the extent to which they protrude beyond the median lobe, can also be diagnostic, as are the relative lengths of the basal piece and the parameres. It should be noted, however, that the median lobe can slide forwards and backwards between the parameres. Pigmentation of the whole structure varies considerably within and between species. The key to *Helophorus* species is followed by a comparison of the aedeagophores (Fig. 101 and Table 1).

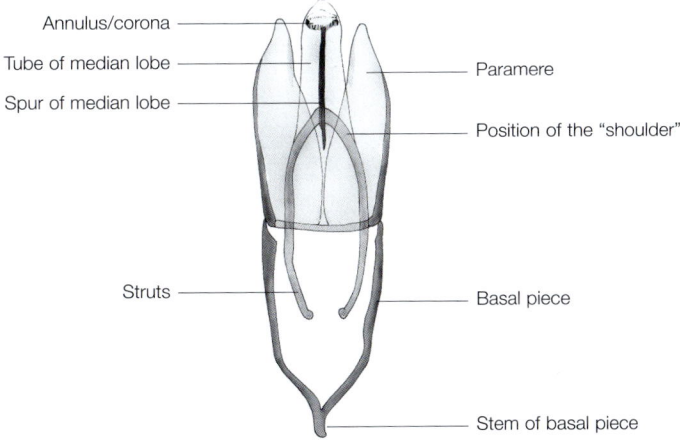

Figure 50. The names used for parts of the *Helophorus* aedeagus

Angus (1992a) should be consulted for all aspects of *Helophorus* in Europe.

Key 2. The species of *Helophorus*

It is unfortunate that some of the commoner species of *Helophorus* are rather nondescript, necessitating some common species to be identified by a process of eliminating rarer species with distinctive features. If a specimen is identified as belonging to one of these rare species, it is advisable to run on through the key to check for better fits among the common species.

1. Body all black with large tubercles (shiny bumps) on the interstices 2, 4, and 6, sometimes also on 8, of the elytra (Fig. 51); median lobe of the aedeagophore markedly shorter than the parameres and with short struts (Fig. 52) .... Subgenus *Kyphohelophorus* Kuwert
4. *Helophorus tuberculatus* Gyllenhal (p. 29)

- Body at least partly brown and without tubercles as above 2

2. Intercalary stria present, a short row of punctures at the front of each elytron in the space between the sutural stria and 2nd stria (Fig. 49) 3

- Intercalary stria absent 8

3. Terrestrial species; inner and middle pronontal intervals strongly raised and protruding in front (Figs 56 and 58); alternate interstices on the elytra developed into keels topped by stiff curved hairs (Fig. 53) .... Subgenus *Empleurus*
4

- Wetland species; front of pronotum straight or gently curved, with inner and middle intervals not protruding in front (Fig. 49); alternate interstices slightly raised but not keeled 6

4. Length 2.8-4.0 mm; elytra with a weak transverse depression just behind the level of the intercalary stria resulting in the ridge of the 2nd interstice being strongly depressed (Fig. 54); aedeagophore about 0.5 mm long, parameres longer than median lobe and strongly narrowed to their tips (Fig. 55) .... **1. *Helophorus nubilus* Fabricius** (p. 28)

The depression (Fig. 54, left) lies about a third of the way down the elytra. This roughly coincides with the first pale area behind the dark triangular mark around the scutellum on those individuals in which the elytral pattern is well developed (Fig. 54, right).

- Length 4.5-5.5 mm; elytra without a transverse depression, the interstices being uniformly raised as ridges; aedeagophore about 1 mm long 5

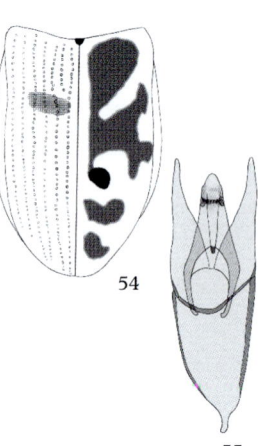

5. Smaller species occasionally up to 5 mm long; front of head with distinctly raised, shining rim; sides of pronotum straight in rear half (Fig. 56); pronotal intervals often strongly narrowed but not broken entirely; shoulders of elytra a little rounded (Fig. 56); parameres curving inwards and slightly shorter than the median lobe (Fig. 57) .... **2. *Helophorus porculus* Bedel** (p. 28)

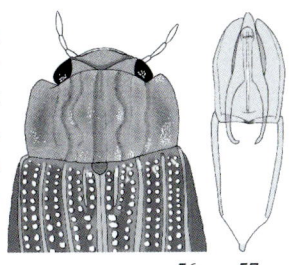

56   57

- Larger species up to 5.5 mm long; front of head without any obvious rim; sides of pronotum sinuate (Fig. 58); pronotal middle intervals broken up by transverse depressions (Fig. 58); shoulders of elytra protruding (Fig 58); parameres more straight and protruding beyond median lobe (Fig. 59) ....
.... **3. *Helophorus rufipes* (Bosc d'Antic)** (p. 28)

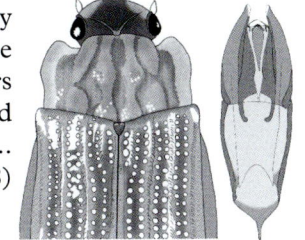

58   59

6. Side of elytron seen from below (the elytral flanks) about as wide as or wider than the epipleur at the level of the hind coxa (Fig. 63); last segment of the maxillary palps symmetrical about its main axis (Fig. 60); hind margin of last abdominal sternite not toothed; aedeagophore with parameres extending much further than median lobe (Fig. 61) .... Subgenus *Trichelophorus*
5. *Helophorus alternans* Gené (p. 29)

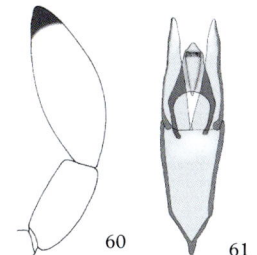

60   61

- Elytral flanks not visible from below or much narrower than the epipleurs at the level of the hind coxa (Fig. 64); last segment of maxillary palps asymmetrical when viewed from some angles, with inner face much straighter than outer face (Fig. 62); hind margin of the last abdominal segment toothed (Figs 67 and 68); aedeagophore with parameres not extending much beyond the median lobe (Figs 65 and 66) .... Subgenus *Helophorus*
7

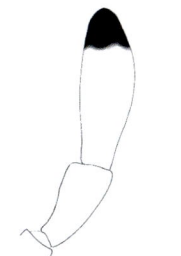

62

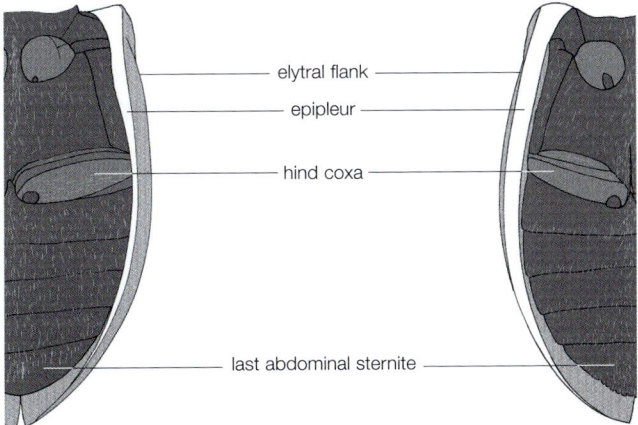

elytral flank

epipleur

hind coxa

last abdominal sternite

Figure 63.
Underside of *Helophorus alternans*.

Figure 64.
Underside of *Helophorus aequalis*

7. "Teeth" of the last abdominal segment small and numerous (14-15 over 0.5 mm – Fig. 67); aedeagophore 0.7-0.9 mm long, parameres not curving out at tip (Fig. 65); smaller (length 4.5-6.3 mm) .... **6. *Helophorus aequalis* Thomson** (p. 29)

- Teeth of the last abdominal segment deeper and distinctly castellated (not more than 10 over 0.5 mm – Fig. 68); aedeagophore 1.1-1.2 mm long, parameres slender, curving out at tip (Fig. 66); larger (size 5.3-7.7 mm) .... **7. *Helophorus grandis* Illiger** (p. 30)

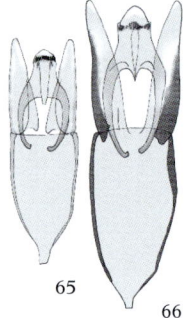

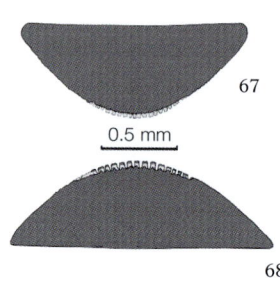

67

0.5 mm

Hair may obscure the view of the abdominal teeth from below. View the last sternite against a white background (× 50 or more): the gap between the elytra may be sufficent to see enough teeth from above. *H. aequalis* and *grandis* are both so common in the spring that it should be possible to put their abdomens side-by-side to demonstrate the clear difference between them.

65

66

68

8. Last segment of maxillary palp almost symmetrical about its main axis viewed from any side, inner and outer faces similarly curved (Figs 69 and 72); sides of elytra (the elytral flanks) very broad when seen from below, at least as wide as the epipleurs at the level of the hind coxae as in Fig. 63 .... Subgenus *Atractohelophorus* Kuwert

9

- Last segment of maxillary palp distinctly asymmetrical, the inner face straighter than the outside face (Figs 75-77); elytral flanks of most species at most as wide as epipleurs as in Fig 64, exceptions with flanks widely visible from below being *H. dorsalis* and *H. strigifrons* .... Subgenus *Rhopalohelophorus* Kuwert

10

9. Last segment of the maxillary palp long, 2½-3 times longer than its maximum width (Fig. 69); sides of pronotum curving in evenly towards hind margin (Fig. 70); aedeagophore with simple parameres (Fig. 71) .... **9. *Helophorus brevipalpis* Bedel** (p. 30)

- Last segment of maxillary palp egg-shaped, its length only about twice its width at the widest point (Fig. 72); pronotum with sinuate sides, curving in then out towards hind margin (Fig. 74); aedeagophore with paramere tips drawn out into points (Fig. 73) ....
.... **10. *Helophorus arvernicus* Mulsant** (p. 30)

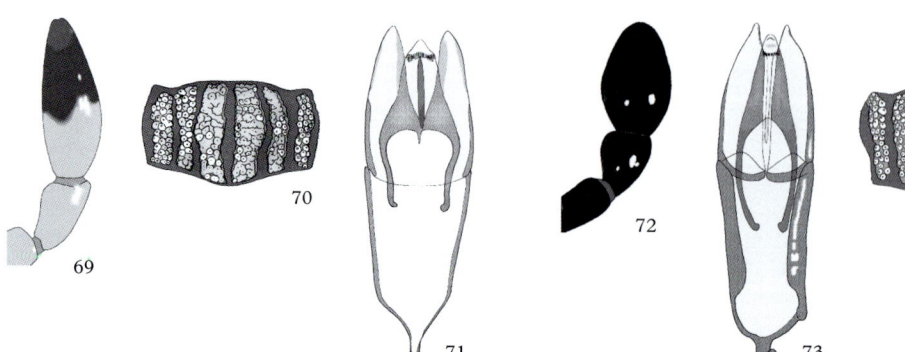

70

69

72

71

74

73

10. Y-shaped groove on the head with a narrow, parallel-sided stem, the front end narrower than the width of either the last segment of the palps or the front tarsi (Fig. 75) 11

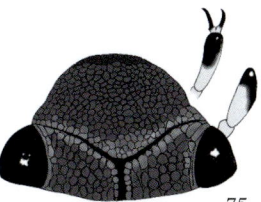

75

-    Y-shaped groove with the stem widening at the front to be about as wide as the width of the last segment of the palps or the front tarsi (Fig. 76) 13

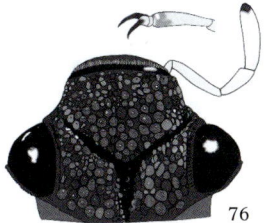

76

11. Head polished except for a few pits, granules being mainly confined to a cluster beside each eye, and usually with an extra groove on either side of the stem of the Y-shaped groove (Fig. 77); pronotum similarly polished and free of granulation; elytral ground colour pale brown, darker markings ranging from elongate smudges either side of the suture in the middle third (Fig. 78) to a much larger mark reaching to just short of the front of the elytra, leaving the rear quarter pale (Fig. 79); aedeagophore Fig. 80 .... **18. *Helophorus nanus* Sturm** (p. 33)

-    Head and most of the pronotum granulate and without extra grooves either side of the Y-shaped groove; elytral ground colour dark brown with at most a few paler spots or, if pale, then with dark marks forming an inverted V, or a pair of rounded spots just past halfway down the elytra (i.e. a more typical *Helophorus* marking) 12

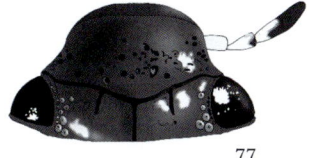

77

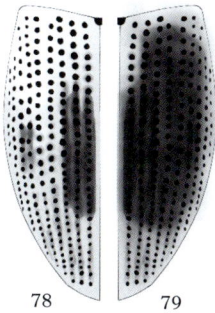

78   79

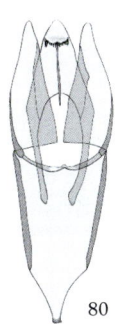

80

12. Elytral flanks as broad as the epipleurs at the level of the hind coxae (Fig. 63); aedeagophore (Fig. 81) with the struts of the median lobe longer than its tube ....
.... **15. *Helophorus strigifrons* Thomson** (p. 34)

-    Elytral flanks not visible from below; aedeagophore (Fig. 82) with the struts of the median lobe about the same length as the tube .... **20. *Helophorus laticollis* Thomson** (p. 33)

A rare species of the New Forest.

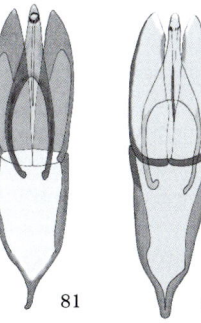

81   82

13. Elytral flanks viewed from below at least half as wide as the epipleurs at the level of the hind coxae 14

-    Elytral flanks hardly, if at all, visible from below 15

14. Elytra dark brown with prominent pale spots (Fig. 83; Plate 10); last segment of palp short (Fig. 84); not in brackish water; aedeagophore with wide parameres (Fig. 85) .... **10. *Helophorus dorsalis* (Marsham)** (p. 31)

- Elytra yellowish, mottled with dark brown; on saltmarshes; last segment of palp long (Fig. 86); aedeagophore Fig. 87 ...... **12. *Helophorus fulgidicollis* Motschulsky** (p. 31)

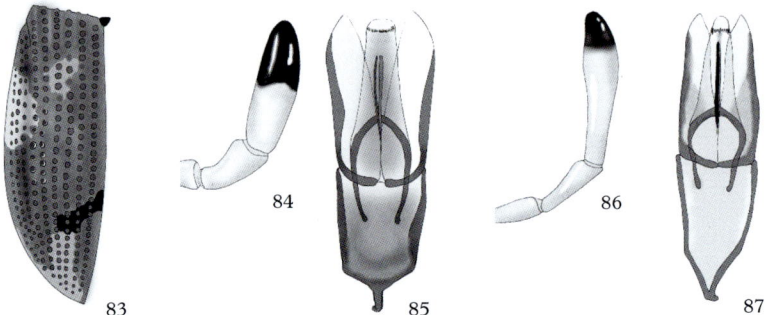

15. A small dark beetle not exceeding 3.0 mm in length; pronotum with a combination of features which are collectively distinct – strongly arched and rounded at sides, all intervals with granules, sometimes those of the internal intervals a little effaced, inner grooves straight or gently curved (Fig. 88); aedeagophore (Fig. 89) very small, about 0.43 mm long .... **13. *Helophorus granularis* (Linnaeus)** (p. 32)

- Length 2.5-4.5 mm; pronotum weakly both arched and rounded, surface with granules effaced in part, inner grooves noticeably sinuate (Fig. 90 *H. minutus*; Fig. 91 *H. obscurus*); aedeagophore 0.48-0.67 mm long 16

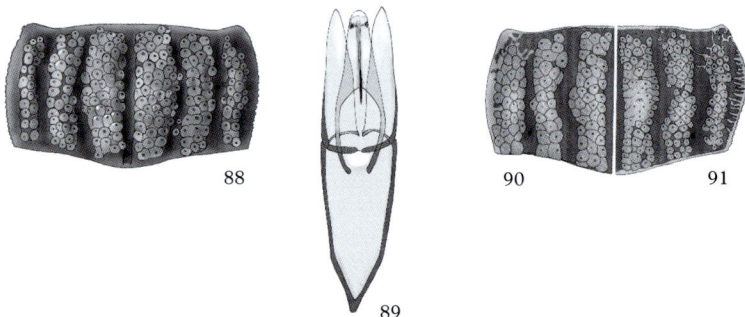

16. Pronotum greenish to golden with front and side margins yellow; elytra yellowish, sometimes mottled, without a V-shaped depression and never with metallic reflections; aedeagophore with struts no longer than the tube of the median lobe (Figs 93, 95 and 97) ... 17

- Pronotum dark brown with front and side margins a dull reddish brown or darker; elytra brown or yellow, with a weak V-shaped depression formed mainly from flattened parts of the 1st to 4th interstices (highlighted in the circle in Fig. 92); elytra often with metallic reflections, sometimes reduced to the bases of the fine punctures on the interstices; aedeagophore with long struts (Figs 98 and 100) 19

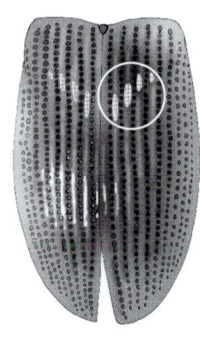

17. Aedeagophore (Fig. 93) with tips of the parameres truncated, the edges sloping inwards but facing more forwards than in other species, and the basal piece longer than the parameres .... **16.** *Helophorus longitarsis* **Wollaston** (p. 33)

- Tips of the parameres pointed and the basal piece about the same length as, or shorter than the parameres (Figs 95 and 97) 18

93

18. Tube of the aedeagophore about the same length as the struts and with shoulders almost absent (Fig. 95); hind corners of pronotum not sinuate (extremes in Fig. 94) .... **17.** *Helophorus minutus* **Fabricius** (p. 33)

- Tube of the aedeagophore longer than the struts and with distinct "shoulders" (Fig. 97); hind corners of pronotum sinuate, at least in the males (extremes in Fig. 96) .... **14.** *Helophorus griseus* **Herbst** (p. 32)

94

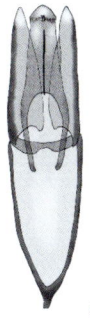

95

96

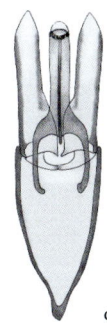

97

19. Aedeagophore pale yellow and small, 0.46-0.60 (rarely 0.65) mm long with a short basal piece (Fig. 98); the parameres splay when a fresh aedeagophore is extruded (Fig. 99) .... **19.** *Helophorus obscurus* **Mulsant** (p. 34)

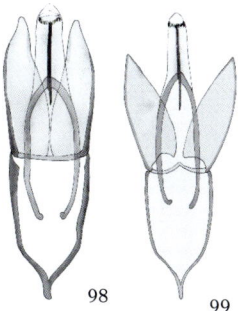

98   99

- Aedeagophore dark brown (unless immature and soft) and larger, between 0.57 and 0.70 mm long with a longer basal piece (Fig. 100) – the parameres do not move apart when the aedeagophore is extruded .... **11.** *Helophorus flavipes* **Fabricius** (p. 31)

Further advice on separating these two common species is given under *H. flavipes* on p. 31.

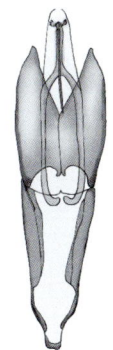

100

Comparison of the aedeagophores of *Helophorus*

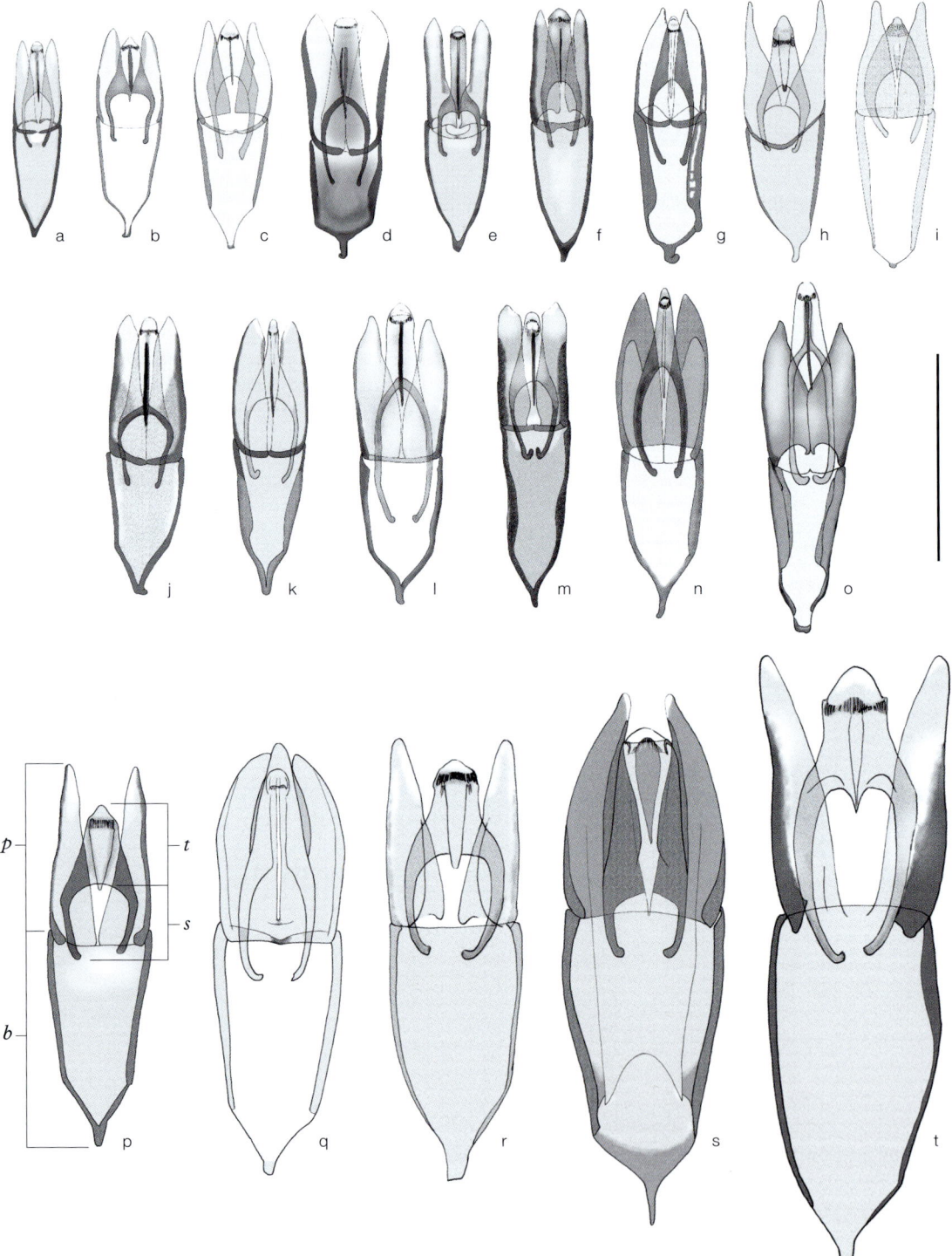

Figure 101 Aedeagophores of *Helophorus* arranged in order of size.
A – *H. granularis*; B – *H. brevipalpis*; C – *H. nanus*; D – *H. dorsalis*; E – *H. griseus*; F – *H. minutus*; G – *H. arvernicus*;
H – *H. nubilus*; I – *H. tuberculatus*; J – *H. fulgidicollis*; K – *H. laticollis*; L – *H. obscurus*; M – *H. longitarsis*; N – *H.strigifrons*;
O – *H. flavipes*; P – *H. alternans*; Q – *H. porculus*; R – *H. aequalis*; S – *H. rufipes*; T – *H. grandis*.

The lengths used to calculate the ratios in Table 1 are on the aedeagophore of *Helophorus alternans* (bottom left): **p** the parameres from their tips to their junction with the basal piece at the side; **b** the basal piece from its junction with the parameres to the extremity of its stem; **t** the tube of the median lobe from its tip to the midpoint; **s** from the midpoint to the tip of the longest strut.

Table 1. Summary of some features of the aedeagophores of British and Irish *Helophorus*. This table is intended to be used in conjunction with Fig. 101.

Species	Length of aedeagophore mm	Ratio of length of struts divided by length of tube of median lobe s/t (see Fig. 101)	Ratio of length of basal piece divided by length of parameres b/p (see Fig. 101)
granularis	0.40	1.1	1.2
brevipalpis	0.41	0.8	1.2
nanus	0.44	1.4	1.3
dorsalis	0.49	1.0	0.9
griseus	0.49	0.5	1.3
minutus	0.50	0.9	1.3
arvernicus	0.50	1.2	1.5
nubilus	0.51	0.9	1.1
tuberculatus	0.53	1.3	1.4
fulgidicollis	0.56	0.7	1.0
laticollis	0.56	1.0	1.3
obscurus	0.60	2.1	1.1
longitarsis	0.61	1.1	1.5
strigifrons	0.66	1.3	1.1
flavipes	0.70	1.9	1.3
alternans	0.75	0.7	1.3
porculus	0.88	0.9	1.5
aequalis	0.90	1.1	1.4
rufipes	1.05	0.4	1.5
grandis	1.20	1.4	1.2

Subgenus *Empleurus* Hope

A most distinctive subgenus with most species confined to relatively dry habitats. These are bulky beetles often having the strongly ornamented upper surface obscured by mud. As might be expected they lack long swimming hairs, having bristles instead. The absence of longer hairs is not used in the keys as these are easily damaged or obscured by glue. A more noticeable feature is that the maxillary palps are sharply pointed.

1. *Helophorus nubilus* **Fabricius** Wheat Shoot Beetle Plate 1

Length 2.8-4.0 mm. *H. nubilus* is unlikely to be confused with any other *Helophorus*, once its usual mud covering is removed. It is small with a hairy, brown pronotum lacking any metallic highlights. The elytra have as a minimum a dark inverted V just beyond halfway to the rear, but more typically have additional spots and brown bars (Fig. 54). The common name is based on the association of larvae with autumn-sown wheat (*Triticum aestivum* Linnaeus). *H. nubilus* is also found in grassland, being taken in pitfall traps, and it appears to be potentially "tecticolous", found living in the *Sedum* matting of the roof of a tower block at Canary Wharf (Jones, 2002). In Spain the species can be found in water in seasonally inundated areas, and there are occasional records from pond margins in Britain and Ireland. *H. nubilus* is recorded over much of lowland Britain but the records are generally old, probably reflecting changes in agricultural practice. The northernmost records are from Speyside. Island records are for the Isle of Man, the Isle of Wight, Skomer and Cardigan Island. Irish records are for Antrim, Armagh, East Donegal, Limerick, Londonderry, Louth, Meath, Waterford, and Wexford. Records are for all months except March, with a peak in September. Being an occasional pest species, the annual cycle of *H. nubilus* is well known, with adults active from May to September, eggs laid in the autumn and larvae active in the winter.

2. *Helophorus porculus* **Bedel** a Turnip Mud Beetle Plate 2

Length 4.0-5.0 mm. The two Turnip Mud Beetles are similar in appearance, with *porculus* being slightly the smaller species with the pronotum more straight-sided and with intervals which are less divided transversely, and with the outer corners of the front of the elytra prominent but rounded. These are the characters usually referred to, but the shining rim to the front of the head appears to provide an additional and obvious character. The association with waterlogged turnips (*Brassica rapa* Linnaeus) and other Brassicaceae is well established but it can also be found on lettuce (*Lactuca sativa* Linnaeus), both adults and larvae being phytophagous. Adults are nocturnal, hiding amongst roots during the day, and the larvae can penetrate deep into the soil in winter. Records include natural habitats on light soils, in particular dune systems. *H. porculus* is poorly recorded recently: its distribution is largely coastal in lowlands across Britain and Ireland, but it would appear to be frequent inland in northern Scotland although it has been recorded from Ben Nevis. Island records are for Great Ganilly in the Scillies, the Isles of Wight and Man, Hayling Island and Mainland Orkney. Adults have been recorded from March to November with a peak in July.

3. *Helophorus rufipes* **(Bosc d'Antic)** a Turnip Mud Beetle Plate 3

Length 4.5-5.5 mm. The largest *Empleurus*, with sharp corners to the elytral shoulders, other features often being hidden by a thick mud coating when it is captured. *H. rufipes* has the same plant and habitat preferences as *H. porculus* on open ground. However, it is decidedly southern in its distribution, being known from England with certainty north to Cheshire and from Wales in Ceredigion and Pembrokeshire. Island records are for Jersey, Tresco in the Scillies, the Isle of Wight and Cardigan Island. Adults have been recorded from April to December with peaks in June and September.

Subgenus *Kyphohelophorus* Kuwert

This subgenus has one extant species and, in North America, two fossil species, all with tuberculate elytra. *H. tuberculatus* is found on moist burnt ground, where it is seemingly camouflaged as burnt fragments.

4. *Helophorus tuberculatus* Gyllenhal The Charcoal Beetle Plate 4

Length 2.8-3.8 mm. The elytral tubercles are so distinctive that this species has been recognised as fragments, in Roman Carlisle and on a rooftop in York. It is often associated with mosses growing on otherwise bare, damp peat and in seasonal pools in burnt areas. *H. tuberculatus* has also been found in tussocks, on a sewage filter bed and swept from rowan blossom (*Sorbus aucuparia* Linnaeus). *H. tuberculatus* is usually found by careful searching but heat extraction in Tullgren funnels has proved effective, as might suction sampling. *H. tuberculatus* has been reported in sufficient numbers to indicate breeding in North Lincolnshire, Derbyshire, South Lancashire, South-east, North-east, South-west and Mid-west Yorkshire in England, and Dumfriesshire and Lanarkshire in Scotland, most of these records being associated with large areas of raised or blanket bog. Other records are for North Somerset, East Sussex, South Essex, and Warwickshire. Recorded from April to November, most records being for April.

Subgenus *Trichelophorus* Kuwert

5. *Helophorus alternans* Gené Plate 5

Length 3.9-5.5 mm. This species can be overlooked as *H. aequalis* in the field, though the dappling of the elytra should draw attention to its being out of the ordinary. *H. alternans* is primarily a species of south-west Europe and North Africa, currently known north to Anglesey. Other Welsh records are for Ceredigion and Meirionydd. Records are almost entirely coastal, in brackish water, but *H. alternans* is also to be found in sun-exposed heathland pools on the Lizard, in the New Forest and in Surrey, the key requirement being warmth. Older records extend the distribution to Scotland on the Solway and to Ireland at Culmore, East Donegal, and in the east to the Wash. Island records are from the Isle of Wight, and from Jersey and Alderney. Records are from March to October with a strong peak in May and June.

Subgenus *Helophorus* Fabricius

6. *Helophorus aequalis* Thomson Plate 6

Length – males 4.5-5.7 mm; females 5.6-6.3 mm. Smaller specimens might be confused with *H. alternans*, but the main problem is the similarity of the larger females to *H. grandis*, necessitating use of the size and number of the teeth on the last visible abdominal sternite. The occasional occurrence of *H. aquaticus* Linnaeus, which is known from Iberia to Denmark on the western side of its distribution, cannot be ruled out in Britain. This would need to be confirmed by examination of chromosomes in freshly prepared material, the slightly smaller size and stronger mottling of *aquaticus* being insufficient to alert the field worker to the possibility of an additional species. In summer *H. aequalis* adults will be found in almost any habitat, but breeding is confined to stagnant freshwater, often impermanent, amongst grasses. This is one of the commonest water beetles, ranging throughout Ireland and Britain, reaching Unst in the north of the Shetlands, but not recorded from much of the Scottish Highlands, and in the Outer Hebrides recorded only from the Uists and Barra. Recorded throughout the year, with a major peak of adult occurrence in June and a minor one in October.

7. *Helophorus grandis* **Illiger** Plate 7

Length – males 5.3-7.1 mm; females 6.1-7.7 mm. This species could only be confused with *H. aequalis*, with which it shares the same habitat preference and often coexists. Common across Ireland and over much of Britain, but scarce north of the Central Belt in Scotland, reaching Westray in the Orkneys, and Lewis and Harris, not being recorded from the Uists. This species starts breeding from early in spring. Recorded throughout the year, with a major peak of adult occurrence in May and a minor one in October.

Subgenus *Atractohelophorus* Kuwert

Angus (1992a) recognised the subgenus *Atracthelophorus* as distinct from *Rhopalhelophorus*, the former being characterised by the symmetrical last segments of the maxillary palps and represented in Britain and Ireland by *H. arvernicus* and *H. brevipalpis*. This treatment is retained here, but with Kuwert's (1886) original spellings – *Atractohelophorus* and *Rhopalohelophorus*.

8. *Helophorus arvernicus* **Mulsant** Plate 8

Length 2.4-3.5 mm. In the field this species is often overlooked as *brevipalpis* but examination of the dark, bulbous terminal segments of the maxillary palps and the strongly domed and highly granulate pronotum should be sufficient to establish identity. The habitat is clean gravel and mud beside running water or on the wave-washed shores of exposed lakes. In Ireland this species is confined to the north, in Antrim, Armagh, East and West Donegal, and Londonderry. *H. arvernicus* is widely distributed on the Scottish mainland, the only island known to be occupied being Arran. The Welsh distribution is mainly in large rivers in Caerfyrddin, Caernarfon, Ceredigion, Denbighshire, Glamorgan, Meirionydd, Monmouthshire, Montgomeryshire, and Radnorshire. In England there is an isolated group of records in West Sussex, Hampshire, Surrey and Oxfordshire, but otherwise *H. arvernicus* is western and northern, reaching southwards as an old record for Leicester and eastwards to Nottinghamshire and South-east Yorkshire. Recorded from March to October, peaking in June.

9. *Helophorus brevipalpis* **Bedel** Plate 9

Length 2.1-4.1 mm. The commonest water beetle over much of Britain and Ireland in the summer, creating problems when making species inventories by sometimes outnumbering all other species and by a bewildering variation in appearance. Though individuals may be found in almost any habitat, breeding only takes place in exposed muddy edges of pools and streams. Apart from the symmetrical last segments of the maxillary palps the pronotum provides the best characters, having a strong metallic finish, the golden grooves contrasting with the intervals, with a granulate surface reflecting red, blue, purple and green; however, the outer margins are narrowly yellow with the extreme edges blackened. The distribution is largely lowland, with *H. brevipalpis* absent from much of highland Scotland, from the Southern Uplands, parts of the Pennines and North Wales. *H. brevipalpis* is recorded north to Yell and Unst in the Shetlands, also to the Faroes and over much of the Hebrides, the Scillies and the Channel Isles. Recorded throughout the year, peaking in July. Parthenogenesis is known in Spain and America.

Subgenus *Rhopalohelophorus* Kuwert

10. *Helophorus dorsalis* (Marsham) Plate 10

Length 3.0-3.8 mm. *H. dorsalis* should be distinguishable in the field from similarly sized *Helophorus* by the pale elytral spots, though *H. fulgidicollis*, in a wholly different habitat, may also be mottled, and other species can have the rearmost pale marks. The habitat is small woodland pools and seeps, often on clay. *H. dorsalis* is known across much of England, but is scarce in old fenland areas. Old records take it north to Kingmoor, Cumberland and Briardene, South Northumberland, but modern records are from North-east Yorkshire southwards. It is largely absent from the south-west, and scarcely reaches Wales, with a single modern record for Monmouthshire and an old one from Glamorgan. Recorded from February to September, and in November, peaking in May.

11. *Helophorus flavipes* Fabricius Plate 11

Length 2.6-4.5 mm. This dark species, with its long and often heavily pigmented aedeagophore, should pose no special identification problems if males are available. The inner grooves of the pronotum have a weaker angle than in *H. obscurus* and the elytra are longer, but females cannot be separated from those of *obscurus* with absolute certainty. *H. flavipes* is more elongate than *H. obscurus* and the rear of the elytra of *obscurus* is more rounded than in *flavipes*; however these characters only become clear with experience, and that same experience will yield exceptions! The habitat is typically shallow water with *Sphagnum*, but almost any acid or circumneutral water can support it. *H. flavipes* is one of the commonest *Helophorus* but has a distinctive distribution being almost entirely absent from the coast of England between Lincolnshire and West Sussex and over much of lowland England apart from on heathland. It ranges to the north of Shetland Mainland, and to St Kilda and the Faroes, and is known from many other islands, but not from the Isle of Wight. It is known from Jersey. Recorded throughout the year, peaking in July.

12. *Helophorus fulgidicollis* Motschulsky Plate 12

Length – males 2.9-3.9 mm; females 4.0-4.7 mm. This is one of the *Helophorus* in which the sexes differ in size enough to give the impression that two species are present. The way in which this species readily swims and dives provides another field character. The translucent orange colour of the pronotum can often be seen through the metallic finish. *H. fulgidicollis* is confined to brackish water, usually in muddy pools with grassy edges in extensive areas of saltmarsh. In Ireland the species occurs in Mid Cork, Down, Dublin, Londonderry, Wexford, and Wicklow, reaching the west coast only in Kerry. In Wales *H. fulgidicollis* is known from Anglesey, Caerfyrddin, Caernarfon, Flint, Glamorgan, Meirionydd, Monmouth, and Pembrokeshire. It occurs from the south coast of England on the Isle of Wight to Moray and East Inverness-shire. *H. fulgidicollis* does not occur in the many small saltmarshes in the west of Scotland where it is known only from the

Solway Marshes. It has been recorded from Guernsey. Recorded throughout the year except January, peaking in April and August.

13. *Helophorus granularis* (Linnaeus) Plate 13

Length 2.2-3.0 mm. This is the smallest of the British and Irish *Helophorus* with the terminal segment of the maxillary palps asymmetrical. The pronotum usually has dark margins, separating such specimens from *H. minutus* and *H. griseus*. The elytra have the interstices uniformly convex as opposed to other species in which the ridging is not uniform or the interstices are flat. The aedeagophore is very small, around 0.43 mm long and marginally smaller than the aedeagophores of the next smallest species, *H. brevipalpis* and *H. nanus*. Finally, most specimens in Britain and Ireland will be found to have shortened wings (Fig. 102), the form *ytenensis* Sharp. The longer wing (Fig. 103) is from a Swedish specimen. The typical habitat is in hard-bottomed pools with fluctuating margins. *H. granularis* is scattered across Britain, from the Lizard to Orkney Mainland. The Irish distribution is in coastal vice-counties, with modern records only for West Mayo. Island records include Clare Island, Arran, Bute and Islay, the Isle of Man, Skomer and the Isle of Wight. Recorded throughout the year, peaking in April with a minor peak in August.

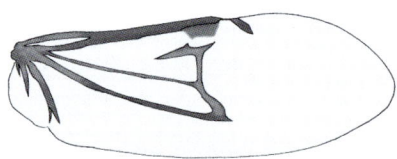

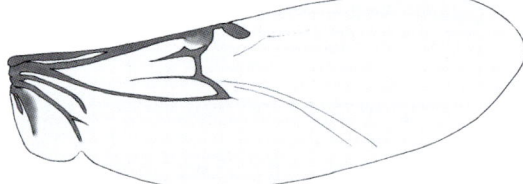

Figure 102. *H. granularis* form *ytenensis* Sharp, shortened winged form typically found in Britain and Ireland.

Figure 103. *H. granularis*, long winged form (from a Swedish specimen).

14. *Helophorus griseus* Herbst Plate 14

Length 2.4-4.0 mm. Apart from the long tube of the aedeagophore the only useful character separating this species from *H. minutus* is the shape of the hind corners of the pronotum. The sides can be weakly sinuate as opposed to straight in *H. minutus*, but this does not apply to all specimens, leaving females to be identified only by association with males. *H. griseus* occurs in similar habitats to *H. minutus*, though perhaps more permanent ones, reflecting the later breeding season. Individuals have been encountered in unlikely places such as mountain pools and a squirrel's drey. Occurrence in saltmarsh conditions, as on Humberside and Teesside, and association with seepages may indicate a requirement for water in midsummer. In Ireland this species is known north to Louth, also from Carlow, Dublin, South-east Galway, North Kerry, Kilkenny, Laoise, Meath, Roscommon, South Tipperary, Waterford, Wexford and Wicklow. This is primarily an English species ranging from West Cornwall to South Northumberland. The Welsh distribution is an extension of the English one, into Caernarfon, Flintshire and Monmouthshire, and the Scottish records are even scarcer, from Fife, Renfrewshire, Roxburghshire and Selkirkshire. The only island record is from the Isle of Wight. Recorded through the year, peaking in June and October.

15. *Helophorus laticollis* Thomson Plate 15

Length 2.9-4.2 mm. In the field this species is likely to be accompanied by *H. flavipes,* from which it may be distinguished by its strongly domed pronotum and, in many specimens, the bipartite colouring, the dark metallic pronotum contrasting with the dull brown elytra. In life it may attract attention as a good swimmer, unlike *flavipes, obscurus* and *strigifrons.* The aedeagophore (Fig. 82) differs from that of *strigifrons* mainly in that the tube is not as long. The inner edges of the parameres are transparent and obliquely cut off at their tips. If slightly splayed, the parameres appear to be straighter than those of *strigifrons* but are in fact unevenly curved on their outer sides. *H. laticollis* is the rarest *Helophorus* in England, currently confined to the New Forest, where it lives in shallowly flooded grass in exposed pools on loam and peat. There are records from heathland in Dorset in the 19th Century and in Surrey up to 1948. Adults have been recorded from March to May, peaking in April, and from September to December. Breeding takes places in winter or early spring.

16. *Helophorus longitarsis* Wollaston Plate 16

Length – males 2.5-3.8 mm; females 3.1-4.5 mm. Under the microscope *H. longitarsis* is a striking animal, though it might be mistaken for a brightly marked *H. minutus* in the field. The pronotum, with flat, lustrous red or pale green middle and inner intervals, is also distinctive in having the band, part transparent and part yellow, on the front edge wider than the median groove. *H. longitarsis* flies readily, has been found in light traps, and is occasionally found as individuals in a variety of habitats. Its breeding habitat is clay or marl bottomed pools that are permanent enough to support it in the summer. These include dew ponds and it can also be found in butyl-lined reservoirs. In Wales it is known from Pembrokeshire and Radnorshire. The English distribution reaches to York and is based on the following vice-counties: South Devon, South Wiltshire, South Hampshire, East and West Kent, East Sussex, Surrey, South Essex, Middlesex, Berkshire, Oxfordshire, East Suffolk, East Norfolk, Northamptonshire, Herefordshire, Worcestershire, Shropshire, Nottinghamshire, Derbyshire, South-east and North-east Yorkshire, and Warwickshire. Recorded from March to October, with most records in August.

17. *Helophorus minutus* Fabricius Plate 17

Length 2.4-3.4 mm. *H. minutus* can be passed over as *H. brevipalpis,* but the asymmetrical terminal segments to the maxillary palps align it with other small pale species, in particular *H. griseus,* the distinction from which is discussed under that species. *H. minutus* is typical of grass-edged water over silt and clay in ponds, lakes, and slow rivers, also occurring in recently cleared drainage dykes and similar habitats in quarries. *H. minutus* is common in central and coastal Ireland. It is also common across most of England, and lowland Wales and Scotland. It is almost absent from the Hebrides and the Highlands, with isolated records for Islay, and Flemington Loch and the Insh Marshes in East Inverness-shire. Other island records are from the Isle of Man, Anglesey, St Mary's in the Scillies and the Isle of Wight. Recorded throughout the year, peaking in May and August.

18. *Helophorus nanus* Sturm Plate 18

Length 2.2-3.0 mm. The unusual elytral pattern of most specimens will set this species apart in the field, the smooth and shining head and pronotum also being distinctive. The larger and rarer *laticollis* can have similar marks at its palest. A more obscure feature unique to *nanus* in the British and Irish fauna is the four-, as opposed to five-, segmented stem to the antenna. *H. nanus* can be abundant in fen conditions amongst grasses and moss. Widely distributed in lowland England, mainly in old fenland areas and on the Brecks, extending north to Westmorland in the west and to Castle Eden Dene, County Durham, in the east. It is absent from south-west England, except for the Somerset Levels (older records from Dorset and South Devon) and it is known in Wales only from the Monmouth Levels. In Ireland recorded from Laoise, Meath, Offaly, and Roscommon. The only record for a small island is from Jersey in 1986. Recorded throughout the year, peaking in April and May.

19. *Helophorus obscurus* Mulsant Plate 19

Length 2.4-4.3 mm. In the field the slightly shorter length and paler colour of most specimens should distinguish *H. obscurus* from *H. flavipes* but such differences can deceive, examination of the aedeagophore being essential for a certain identification. The way in which the parameres splay apart when pressure is applied to a recently killed specimen is distinctive. The basal piece is also shorter than that of *H. flavipes*. *H. obscurus* lives in muddy situations, from neutral and alkaline waters. *H. obscurus* is frequent through lowland England, Wales and Scotland but is scarcer in Ireland. It reaches Eday in the Orkneys and in Scotland is also reported from Arran, Barra, Bute, Cumbrae, Islay, Jura, Mull, and Skye. Other islands with records include the Isle of Man, Jersey, St. Mary's in the Scillies, Sherkin and Saltee. Recorded throughout the year, peaking in April and October.

20. *Helophorus strigifrons* Thomson Plate 20

Length 3.2-4.2 mm. The best dorsal character separating this species from the similarly sized *dorsalis*, *flavipes* and *obscurus* is the narrow stem to the Y-shaped groove on the head. This beetle is often caked in mud and its surface may only be seen after application of a strong detergent, cleaning out crevices with a short-haired brush. The elytral flanks are widely visible below, separating this species from *flavipes* and *obscurus*, and also from *laticollis*. The aedeagophore has shorter struts than in *flavipes* and *obscurus* and the whole structure is typically a rich brown colour, the rounded outer faces of the parameres also being distinctive. The whole insect is also typically dark brown but a pale form is known from England and Scotland. *H. strigifrons* is associated with temporary marshes with rush and sedge litter, usually in well-established sites, and can be caught in pitfall traps and by suction sampling in wet turf. *H. strigifrons* is widespread on low ground throughout mainland Britain, the only island records being from Anglesey and Cumbrae. In Ireland the few records are mainly from the south in Clare and South-east Galway in turloughs, also from Kerry, Waterford and Wicklow. Recorded throughout the year, peaking in April and September.

Family GEORISSIDAE Castelnau

1. *GEORISSUS* Latreille

These are known as mud-loving beetles. Once any adhering soil particles have been removed there can be no doubt about the identity of this genus, of which only one species is frequent across northern Europe. There are many distinguishing features, the cryptic habit, the body shape, and the four-segmented tarsi being more than enough to set the sole British and Irish species apart.

1. *Georissus crenulatus* (Rossi) Plate 21

Length 1.4-2.1 mm. This small black species is difficult to detect alive except in strong sunlight when the occasional movement of what appears to be a mud particle may attract attention. The crinkly "shelf" on the front of the pronotum distinguishes *G. crenulatus* from other European species. Occasionally a specimen has the median groove of this sculpturing running two-thirds of the way over the pronotum but the main part of the pronotum is typically smooth and shining. If the area thought to be occupied by the beetle is deliberately flooded specimens will float just below the surface or fall to the bottom, having little hydrofuge hair and often being ballasted by adhering soil particles. *G. crenulatus* can also be caught and detected using pitfall traps, as has been done on exposed riverine sediments. The habitat, on moist soil or sand, has a crumbly texture with a biofilm of algae or of fine mosses, a habitat found on the banks of water bodies, beside cliff seeps and in duneslacks. *G. crenulatus* is known from Ireland in Roscommon, Clare and North Kerry. It is scattered across Britain being known north to St. Cyrus in Kincardineshire and the River Tummel in East Perthshire. It appears to be absent from much of the English Midlands, also from south-east England, and at its most common in Wales, with records for six vice-counties. Islands recorded are Islay, the Isle of Man, Anglesey and the Isle of Wight. Recorded from April to September with a strong peak in June, also reported in February and November.

Family HYDROCHIDAE Thomson

1. *HYDROCHUS* Leach

This is another distinctive genus, with a shining black or metallic, heavily ornamented, ridged, pitted and dented upper surface, and a black velvety finish to the underside, which has deep pits. The body shape is elongate though two species have the rear end characteristically flared. Features associated with the rows of large punctures (referred to as punctured striae) are important in identification, in particular the interstices that lie between them, which may be partly ridged. These interstices each have a series of very fine punctures visible at × 50. The rear of the elytra may have a few very large punctures: these may be partly hidden under a ledge, requiring the beetle to be tilted to give a better view.

The sexes can usually be distinguished by the last sclerotised abdominal sternite (the eighth) being broader in females than in males. Males should be routinely dissected. The last sclerotised sternite can provide characters to separate females of different species; it should not be confused with a semitransparent lamella that protrudes beyond it. The aedeagophore is divided into a basal piece, large in two species and small in the rest, and two parameres flanking the median piece which may have a threadlike "flagellum" at its tip. One paramere has an expanded tip in some species. That paramere is anatomically the left one but appears on the right in the figures as these are dorsal views.

The last critical review of European *Hydrochus* was by Angus (1977) but one northern European species has been subsequently recognised (van Berge Henegouwen, 1988) and *H. carinatus* Germar is now known as *H. crenatus*.

Key 3. The species of *Hydrochus*

1. The interstices between alternate punctured striae on the elytra raised into sharp ridges, so that the striae are in pairs (Fig. 104) – the ridges can be complete or incomplete 2

- Interstices equally raised, except in one species with slight ridges on the front edge and near the middle of the elytra, between the 4th and 5th striae 6

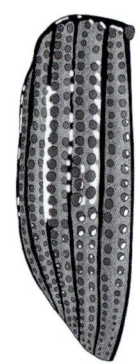

104

2. Ridge between the 2nd and 3rd striae ending halfway back, and a ridge between the 3rd and 4th striae beginning close to this point (Fig. 105) 3

- Ridge between the 2nd and 3rd rows of striae running the full length of the elytra 4

105

3. Space between the 3rd and 4th striae with a short ridge at the front edge of the elytra, about same width as the neighbouring ridges (Fig. 107); male genitalia with the tip of one paramere greatly enlarged and with a slot in it to receive the other one (Fig. 106); female with a notch on either side of the last sclerotised abdominal segment (Fig. 108) ....
.... **4. *Hydrochus elongatus* (Schaller)** (p. 39)

- No ridge in the position described above (Fig. 110), or at most a very weak one; male with tip of one paramere only slightly enlarged (Fig. 109); sides of the last sclerotised abdominal segment of the female simple (Fig. 111) ....
.... **5. *Hydrochus ignicollis* Motschulsky** (p. 39)

107

106

108

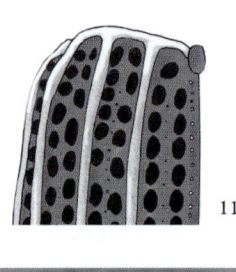

110

109

111

4. Body elongate, elytra almost twice as long as the total width; large elongate punctures on hind edge of the elytra, the largest running across the bases of the 1st and 2nd striae (Fig. 112); aedeagophore characteristic, with parameres longer than the basal piece and separate from the median lobe, which has a short flagellum (Fig. 113) ... **3. *Hydrochus crenatus* (Fabricius)** (p. 38)

- Body short and broad, elytra about 1½ × as long as wide; hind edge of the elytra without especially large punctures (Fig. 114); basal piece of the aedeagophore longer than the parameres (Figs 117 and 118), which hide the median lobe 5

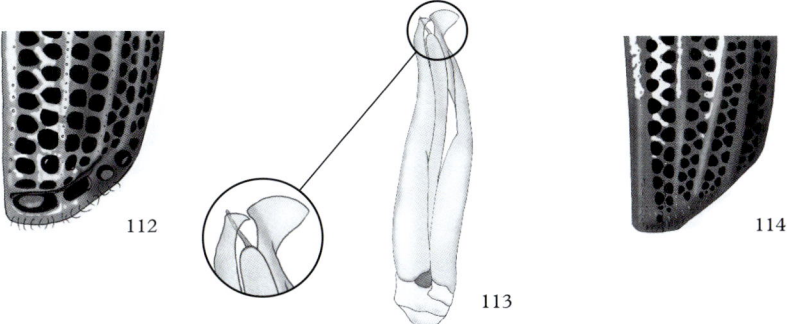

5. Aedeagophore small with rear end of basal piece open to receive sperm duct (Fig. 117); pronotum broadest near to the front (Fig. 115) ... **6. *Hydrochus brevis* (Herbst)** (p. 38)

- Aedeagophore strikingly large, ⅓ of body length, and basal piece with duct entering it from the side at its rear (Fig. 118); pronotum broadest nearer to the middle than in *H. brevis* (Fig. 116) .... **6. *Hydrochus megaphallus* van Berge Henegouwen** (p. 39)

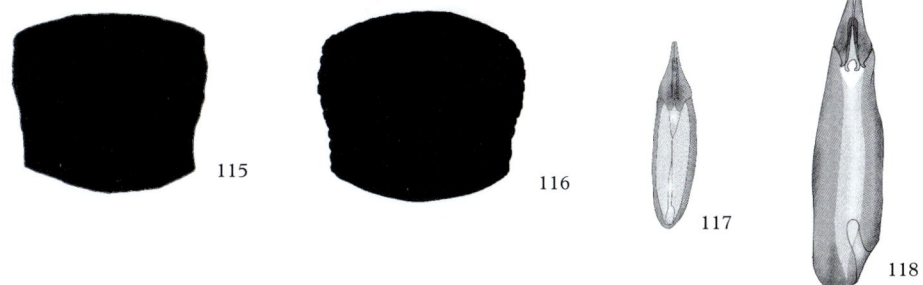

6. Smaller species, up to 3 mm long; maxillary palps dark throughout (Fig. 119); legs with broadly dark bands around the joints, essentially "black knees" visible even with a hand lens; aedeagophore with the parameres similar to each other, being narrowly tipped; median lobe with a flagellum (Fig. 120) .... **7. *Hydrochus nitidicollis* Mulsant** (p. 39)

- Larger species, 3 mm or more in length; maxillary palps darkened at tips (Fig. 121); leg joint darkening more diffuse; parameres dissimilar, the left one having a broad tip; median lobe without a flagellum (Fig. 122) .... **1. *Hydrochus angustatus* Germar** (p. 38)

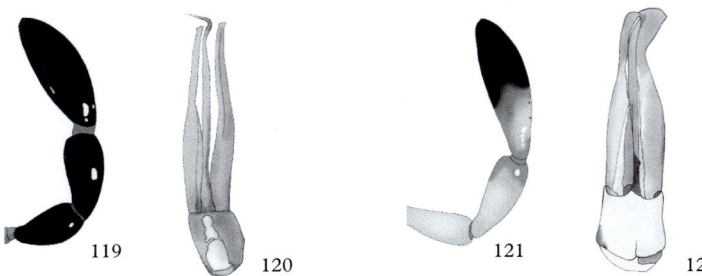

1. *Hydrochus angustatus* Germar Plate 22

Length 3.0-4.0 mm. This slender species, with its elytra lacking any lengthy ridges, ought not to be confused with any other species but it has been mistaken for *H. nitidicollis*, an even more localised species. Both it and *H. nitidicollis* may have short ridges, no longer than the space occupied by five of the elytral punctures, in the rear half of the elytra between the 4th and 5th striae, and they also have weak signs of ridges elsewhere. *H. angustatus* is larger than *H. nitidicollis*, and its elytra are proportionately longer (more than twice their width at the widest point, as opposed to less than this in *H. nitidicollis*). Given the misidentification problems that have been encountered it is safest to examine the aedeagophore to confirm the species involved. A potentially British species reaching northern France is *H. flavipennis* Küster: it resembles *H. angustatus* in having the median lobe lacking a flagellum but its left paramere is less expanded at the apex. *H. angustatus* lives in a wide range of lowland, permanent still waters with some exposed mud or peat, also sometimes in the vegetated edges of rivers. It is typical of heathland ponds in southern England. In Scotland it was known only from the Maxwelltown Loch in 1892, an area now drained and lying in a built-up part of Dumfries, but there are several records from the Solway Merse on the Cumberland side. It is absent from the east coast north of the Wash, and is scattered elsewhere in England and Wales, with most inland records being very old, as are those from Anglesey and Jersey, but there are both old and new records from the Isle of Man and Guernsey. In Ireland there are old records from Limerick, Waterford and Wexford. Recorded throughout the year, peaking in April and September.

2. *Hydrochus brevis* (Herbst) Plate 23

Length 2.8-3.7 mm. *H. brevis* can only be confused with the much rarer *H. megaphallus*. External features that might differentiate the species are not reliable without reference to the aedeagophore, but they might at least emphasise the need for dissection. The pronotum varies in shape, with that of *brevis* typically being widest very near the front, whereas it is widest nearer to the middle in *megaphallus*. The last visible abdominal sternite is typically longer than in *megaphallus* but there are exceptions. *H. brevis* is largely confined to mesotrophic and acid fens in a scattered distribution based on ancient fenland. In Ireland it is found in cutover bog and similar habitats in Antrim, Armagh, Fermanagh and Leitrim and in Wales it is known only from Anglesey. In Scotland it occurs in the south in Berwickshire, Kirkcudbrightshire, Roxburghshire and Wigtownshire, and in the north in Angus and on Speyside in East Inverness-shire and Moray, with old records from Dumfriesshire, Midlothian, Mid Perthshire, and Renfrewshire. The Brecks and the Broads provide the stronghold for this species in England, with other populations in County Durham, Cumberland, Shropshire, and Mid-west and South-east Yorkshire. Recorded in all months except January, peaking in April and September.

3. *Hydrochus crenatus* (Fabricius) Plate 24

Length 2.1-3.1 mm. The narrow body form and well developed continuous ridges on the elytra would be enough, with care, to distinguish this species from the rest. This is a fen species, particularly associated with fluctuating meres with much moss in their drawdown zones. *H. crenatus* is largely confined to East Anglia, particularly on the Brecks and around Woodwalton Fen in Huntingdonshire. It extends to Potteric Carr in South-west Yorkshire, the Burton pits in North Lincolnshire, and to Flatford Mill in East Suffolk. A proposed old record by the Reverend Little (Dobson *et al.*, 2012) for Scotland is unlikely to be correct. There are subfossil records from North Somerset and recent ones at Dungeness, East Kent in a previously well-worked area indicating colonisation from 2000 onwards. Recorded throughout the year, peaking in April/May and September.

4. *Hydrochus elongatus* **(Schaller)** Plate 25

Length 3.2-4.7 mm. This is marginally the largest *Hydrochus* species in Britain, only to be confused with *H. ignicollis* and recognised by the short additional ridges on the elytra and by the male genitalia. Shaarawi & Angus (1992) noted a triploid female embryo in this species in their studies of hydrophiloid karyotypes: such triploids occasionally give rise to parthenogenetic races, as in some *Helophorus brevipalpis* (Angus, 1992b), and the possibility of this phenomenon should be borne in mind for this species. *H. elongatus* is mainly found in base-rich fens, often reedbeds, in lowland England, particularly in the east from Humberside southwards and also in the Severn Valley, its distribution including most Midland vice-counties. There is an old record for Glamorgan, one requiring confirmation from Midlothian, and old records in England indicating a wider distribution in the past. Recorded in all months except January, peaking in May and August.

5. *Hydrochus ignicollis* **Motschulsky** Plate 26

Length 3.4-4.1 mm. This species is only likely to be confused with *H. elongatus*. The typical habitat is mesotrophic fen amongst moss or, in Ireland, lake shore fens with much reed or sedge litter. The distribution is unusual with more sites known for it in Ireland than in England. The Irish distribution is central, extending out to Limerick, the Burren and Antrim. In East Anglia the species is frequent in lithopalsa fens in West Norfolk and in the Catfield area of East Norfolk, at Wicken Fen in Cambridgeshire, and at a few sites in South Essex, Northamptonshire and South Lincolnshire. Other than Anglesey, where the only two known Welsh sites occur, and the Somerset Levels, all the other records are eastern, in Buckinghamshire, East and West Kent, and East Sussex. Recorded in all months except December, peaking in August.

6. *Hydrochus megaphallus* **van Berge Henegouwen** Plate 27

Length 2.7-3.3 mm. This species can be found in association with *H. brevis*, but it is occasionally common on its own in more base-rich dykes and broads. Its distinction from *H. brevis* is discussed under that species. *H. megaphallus* is confined to the Broads in East Norfolk at Catfield, Hickling and Burgh Common and in East Suffolk in Redgrave and Lopham Fens. There is an old record from Leicestershire. Recorded from March to July and from September and October, peaking in May.

7. *Hydrochus nitidicollis* **Mulsant** Plate 28

Length 2.4-3.0 mm. This species has been confused with *H. angustatus* in the past, and dissection is advised for any suspected specimens, particularly if found outside the known range in south-west England. The habitat is shallow water over an exposed substratum of mud such as china clay or partly silted beds of riverine sediment. The only authenticated records are from East and West Cornwall, and South and North Devon. Recorded from April to September, with a strong peak in September.

Family SPERCHEIDAE Erichson

1. *Spercheus* Kugelann

The sole Palaearctic species of this genus could not be mistaken for any other beetle were it not that it is sometimes so covered in mud that it has a passing resemblance to a similarly encrusted aphodiine dung beetle or to a large *Helophorus*. The adult is easily recognised (Fig. 123) without resorting to family characteristics such as the antennae being seven-segmented but appearing six-segmented, the 3rd segment being minute, and the abdomen having five visible sternites. As in *Helochares*, the females carry the egg sac.

1. *Spercheus emarginatus* (Schaller) Plate 29

Length 5.5-7.0 mm. The adult beetle, once cleaned, has a distinctive elytral pattern that is usually asymmetrical (Fig. 123). The head has a wide median notch, the males having the anterior corners of the clypeus distinctly angled (Fig. 123), whereas these are rounded in females (Fig. 124). When disturbed the larvae float just below the surface and are also distinctive, being of the hydrophiloid type but bulbous, with the head and legs blackened, and wart-like processes on the sides of most abdominal segments. Their presence is often detected before catching an adult. The habitat is muddy ponds and ditches in fenland, usually with rich vegetation. The last British record was in Beccles, East Suffolk, in 1956. All other records are from the 19th Century in South Essex, Middlesex, Berkshire, Cambridgeshire, Huntingdonshire, and mid-west Yorkshire. These include the earliest English records of any hydrophilid beetle (Leach, 1817) from near York and as elytral fragments in Kensington Gardens. English records are from July and November: in western Europe adults become active in the spring or summer, with rapid development of the larvae from the egg sacs held by the females, several broods being produced, and newly emerged adults being found from the middle of June.

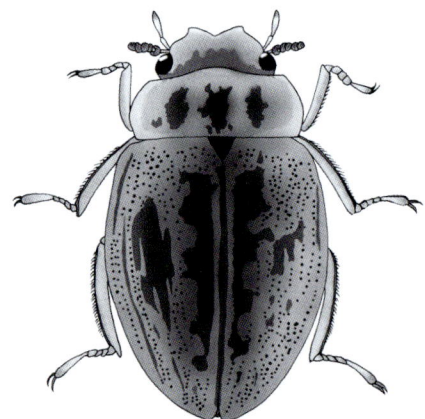

Figure 124. *Spercheus emarginatus*, head of adult female.

Figure 123. *Spercheus emarginatus*, adult male.

Family HYDROPHILIDAE Latreille

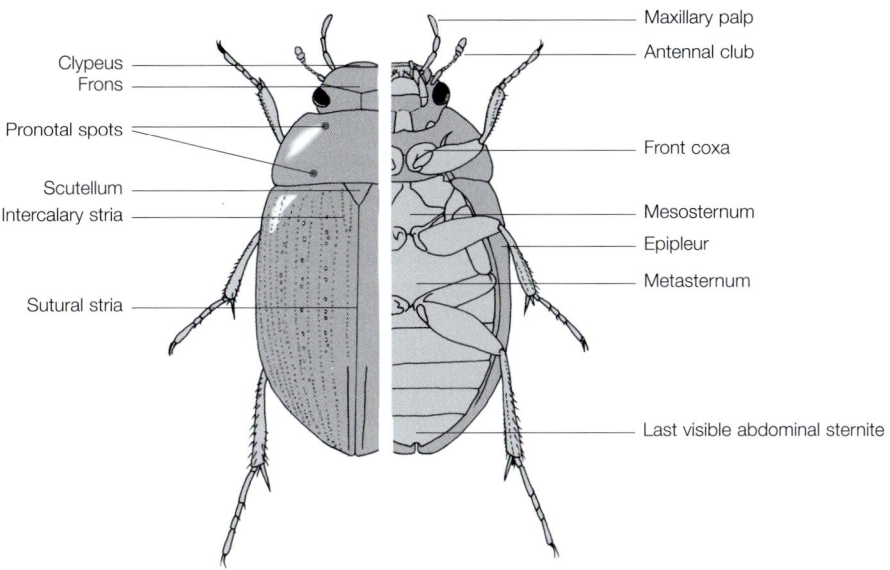

Figure 125. A generalised hydrophilid, labelled to show parts used in identification.

Members of the family Hydrophilidae, as defined here, are typically convex and with a more or less continuous body outline viewed from above, the pronotum being wider than long and widest at the rear. There is a technical character unlikely to be used in identification, in that the cavities in which the front coxae swivel are interconnected at their rear. The major subdivision into the subfamilies Hydrophilinae and Sphaeridiinae is also based on relatively obscure characters. The upper lip or labrum of Hydrophilinae is strongly sclerotised and not paler than the clypeus, whereas it is weak and pale in Sphaeridiinae and often retracted under the clypeus. The sphaeridiine labial palps are very small with more hair than in hydrophilines. What appears to be the basal segment of the maxillary palp (actually the second) is usually broader than the other two visible segments. A much more obvious feature is that Sphaeridiinae have compact antennal clubs that are prominent in life (Plates 69-102) with the antennae being almost as long as the maxillary palps. The Hydrophilinae are more variable in the relative lengths of the palps and antennae (Plates 30-68), the clubs also being variable, looser than in Sphaeridiinae and often hidden beneath the head in life. Ecologically the major difference is that all of the British and Irish Hydrophilinae are aquatic whereas most Sphaeridiinae are terrestrial. Hansen (1987) also drew attention to the possibility that many Hydrophilinae have one generation a year whereas several Sphaeridiinae have two. It is, however, possible for some Hydrophilinae to have several broods within a year.

The genera of Hydrophilidae are keyed along with other members of the Hydrophiloidea and with the Hydraenidae (Key 1, page 10) as "palpicorns", rather than within families.

1. *ANACAENA* Thomson

These are small convex species characterised by the elytra lacking any lines of punctures or grooves other than the sutural striae, which run from about halfway to the apex. The colour is brown or black and never metallic. The underside has much water-repellent pubescence, the distribution of this on the hind femora being important in distinguishing species and thus requiring that some carded material should be mounted sideways or dorsal side down with the underside free from glue. The process on the mesosternum is also important. The great variability recorded in this genus has been resolved in western Europe with the recognition by van Berge Henegouwen (1986) that two species had been lumped together as *A. limbata*. The sexes cannot usually be distinguished unless the extremities of the genitalia are protruding; however, the end segment of the fore tarsus is expanded in the male of *A. lutescens*. The aedeagophore does not offer any critical identification features but the geographical distribution of the sexes is of interest in *A. lutescens*.

Key 4. The species of *Anacaena*

1. A small species, 2.5 mm or less in length, with the upper surface predominantly yellow, in particular the patch in front of each eye (Fig. 126); hind femora with the hairline scarcely reaching the hind edge (Fig. 127) ....
.... **1. *Anacaena bipustulata* (Marsham)** (p. 43)

- Larger species, 2.4 mm or more in length; body darker with area in front of eye reddish brown at its most pale; hind femora with hair line reaching at least a third of the rear edge (Figs 131, 133 and 134) 2

126   127   128   129

2. Body shape decidedly globular (Fig. 128); upper surface black unless immature, the outer margins of the pronotum and elytra paler with the colour gradually intensifying towards the middle, elytra never with a spot centred on the suture; mesosternum just above the level of the middle coxae with at most a transverse ledge (Fig. 130) barely protruding when viewed from the side, never with a protuberance; hind femora with about a third of the surface lacking pubescence, the hairs running three-quarters of the way along the front edge and less than a third of along the rear edge (Fig. 131) ....
.... **2. *Anacaena globulus* (Paykull)** (p. 43)

- Body shape more narrow as in Fig. 126; upper surface black or uniformly brown, often with a large dark spot on the suture (Fig. 129); mesosternum with a longitudinal keel leading onto a tooth-like backwardly pointing process in the midline just in front of the mid-coxae (Fig. 132, a – viewed from below and b – from the side); hind femora with the hair line reaching much of the rear margin (Figs 133 and 134) 3

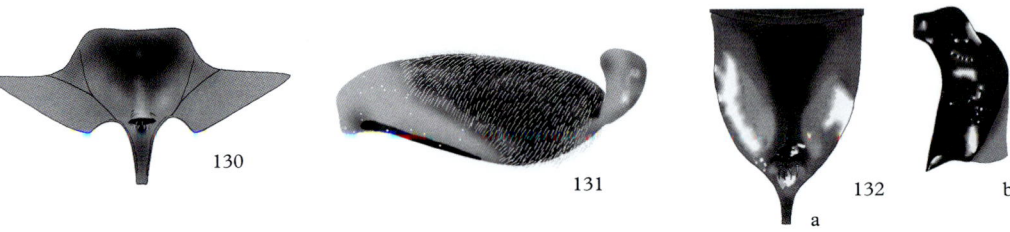

130   131   132   b
a

3. Felt patch on the hind femur covering almost the entire ventral face, extending more than ¾ the way along the hind margin (Fig. 133), on which a sliver without hairs can sometimes be detected .... **3. _Anacaena limbata_ (Fabricius) (below)**

\- Felt patch on hind femur broadly cut away on hind margin, not reaching more than half way along it (Fig. 134) .... **4. _Anacaena lutescens_ (Stephens) (p. 44)**

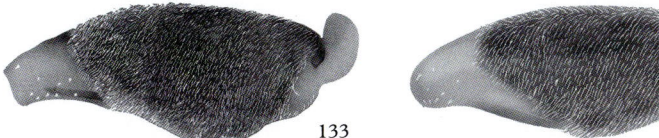

133   134

1. _Anacaena bipustulata_ (Marsham) Plate 30

Length 1.9-2.5 mm. The small size and pale, dappled form are distinctive in the field. Although not used in the key it might be noted that the mesosternum has a small sharp longitudinal keel unlike the transverse ledge in _A. globulus_ and the keel leading to a pointed process in _A. limbata_ and _A. lutescens_. _A. bipustulata_ is mainly associated with slow running water in lowland, particularly on a partly exposed clayey substratum in grazing fen ditches, canals, quarry ponds and similar manmade habitats, sometimes in weakly saline areas. This species is common on the east side of England up to south-east Yorkshire, but more sparsely distributed in the west, the most northerly record on the coast in the west being for Margam Moors in Glamorgan. _A. bipustulata_ is known from Guernsey and Jersey. Recorded in all months of the year except December, with strong peaks in May and September.

2. _Anacaena globulus_ (Paykull) Plate 31

Length 2.5-3.0 mm. The more rounded shape and darker colouring should set _A. globulus_ apart from the other species. If in doubt an elytron should be lifted to show the gradation from its dark centre to the paler periphery, and the absence of a sutural spot. The weak ledge in front of a sensory pit on the mesosternum is also distinctive but difficult to see. _A. globulus_ is almost ubiquitous in any damp habitat except saltmarsh across much of Britain and Ireland, ranging north to St Kilda and Unst. It is not recorded from the Scilly Isles but occurs on Jersey. It is generally commonest in the edges of running water but in the extreme north of its range, in the Orkneys, Shetland and the Faroes, it is common in the mossy edges of permanent still water, a habitat occupied by _A. lutescens_ and _Enochrus affinis_ elsewhere. There are gaps in its distribution in England largely on the chalk. _A. globulus_ is the most globular of the British and Irish species but in the south of its mainland Europe range it is almost as narrow as the others. This may well reflect a difference in flight muscle development, British specimens having originally been reported as incapable of flight by Jackson (1973). Frequently recorded throughout the year, with peaks in June and August.

3. _Anacaena limbata_ (Fabricius) Plate 32

Length 2.4-3.2 mm. The extent of the fur patch on the hind femur provides the only safe method of separating this species from _A. lutescens_ and even this requires careful manipulation of light to view the rear edge of the femur. Forms with a weakly pale area in front of the eyes should prove to be _A. limbata_ rather than _A. lutescens_, but a decidedly pale area is found only in *A. bipustulata*. *A. limbata* appears to be a very variable species with a diminutive form often abundant on coastal levels. The habitat is typically well-vegetated, mostly eutrophic, still waters, but *A. limbata* can coexist with *A. lutescens* in peaty habitats. *A. limbata* is common across much of lowland England and coastal Wales, reaching north to County Durham in the east and single sites known in Westmorland, Cumberland, Dumfriesshire and Wigtownshire. There are also records for the Isle of Man, Lundy, Sherkin Island, St. Mary's and Tresco in the Scillies, the Isle of Wight and Guernsey. Irish records are mainly from the midland vice-counties. Recorded throughout the year, with peaks in May and August.

4. *Anacaena lutescens* (Stephens) Plate 33

Length 2.5-3.2 mm. The extent of the felt patch on the hind femora is critical in identifying this species. *A. lutescens* is unusual in that the claw-bearing final segment of the front tarsi is slightly expanded in the male: the character is too difficult to see to provide a useful distinction from *A. limbata* and, as indicated below, males are not necessarily present in populations. *A. lutescens* is frequent across much of Ireland and Britain north to Caithness, living in well-vegetated still waters, also amongst *Sphagnum*. It can also occur in woodland pools amongst dead leaves with *A. globulus*. It occurs in most of the Hebrides, Barra, Islay, Jura, Mull, Raasay, South Rona, Skye, Soay, and South Uist. *A. lutescens* appears to be unique in the British water beetle fauna in having a parthenogenetic population associated with diploidy and triploidy, one of the nine chromosomal pairs being heterozygous (Shaarawi & Angus, 1991). Males have been found in Ireland, the Isle of Man, Wales and southern England. The parthenogenetic form was first reported in Cumberland and is the only form so far known in Scotland. In contrast the forms varying in colour and elytral spotting occur throughout the species' range, with the dark form more frequent in acid and upland sites. Recorded throughout the year, peaking in April and May.

2. *PARACYMUS* Thomson

The two species are similar in size and shape to *Anacaena*, from which they should easily be distinguished by the entirely dark body, sometimes a little paler at the edges, and the shining hind femora. The body has metallic tints and is shining even though it is coarsely punctured. As in *Anacaena* the only elytral striae are those either side of the suture; they are longer than in *Anacaena*, reaching nearly to the scutellum. The antennae are often claimed to have only eight segments, in contrast to the nine in *Anacaena*, but the species in Britain and Ireland currently understood to be *P. scutellaris* (Rosenhauer) has nine segments, the extra one being amongst the very small segments below the cupule of the club. The claw-bearing segment of the front tarsus is slightly dilated in males and spinier than in females. Although it should not be necessary to examine the genitalia the aedeagophores of the two species differ markedly even at their extremities, which are often exposed without dissection.

Key 5. The species of *Paracymus*

1. Maxillary palps yellow except for darkened outer half of terminal segment (Fig. 135); elongate oval body outline, with parts of elytral sides straight (Fig. 136); mid femora with the inner third hairy (Fig. 137); aedeagophore (Fig. 138) distinctive, with the pointed parameres having two flaps .... **1. *Paracymus aeneus* (Germar)** (below)

- Maxillary palps darkened throughout with at least the outer two-thirds of the terminal segment black (Fig. 139); short oval body outline with elytral sides rounded throughout (Fig. 140); mid femora two-thirds hairy (Fig. 141); aedeagophore distinctive, with bulbous and transparent tips to the parameres (Fig. 142) ....
.... **2. *Paracymus scutellaris* (Rosenhauer)** (below)

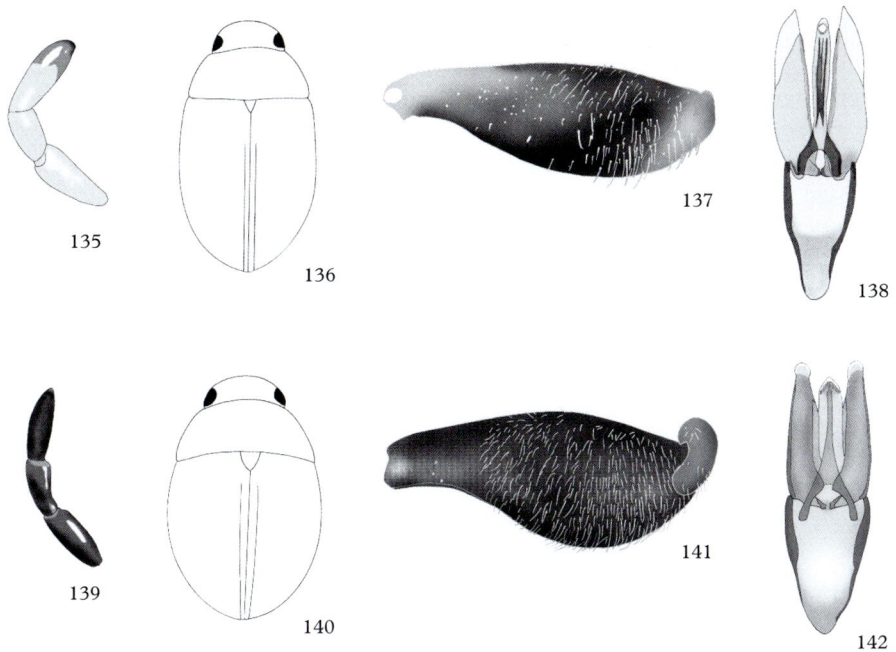

135
136
137
138
139
140
141
142

1. *Paracymus aeneus* (Germar) Plate 34

Length 2.5-3.5 mm. The beetle itself can easily be separated from *P. scutellaris*. No other small black and globular species is likely to be found in a saltmarsh, where it can be common over mud but is more usually found in the shelter of marginal vegetation. *P. aeneus* is known from the Isle of Wight, and in South and North Essex. Recorded from April to November with peaks in May/June and in September.

2. *Paracymus scutellaris* (Rosenhauer) Plate 35

Length 2.5-3.2 mm. This species can be passed over as *Anacaena globulus* in poor light in the field, but the metallic sheen is otherwise a good field character. The habitat is typically shallow seepage over peat, often amongst mosses, but single specimens can be found in other wetlands. The distribution is generally western and coastal with some exceptions. In Ireland *P. scutellaris* is widely distributed, being recorded from Antrim, West Cork, West Donegal, County Down, Fermanagh, West Galway, North and South Kerry, Laoise, Longford, Offaly, Sligo, Waterford, Westmeath, and Wexford, including several islands –

Achill, Clare, Inishbofin, Rathlin and Saltee. In Scotland *P. scutellaris* ranges from the Mull of Kintyre to Caithness, being mostly found on islands – Barra, Bute, Coll, Fladda, Islay, Jura, Lewis and Harris, Raasay, South Rona, Rum, Scalpay, Skye, Tiree, and South Uist. This species is very rare in northern England, with single records from Cumberland, and North-east and South-east Yorkshire, the latter from Spurn. In the southern half of England the beetle occurs mainly in the south-west, extending to Warwickshire, and on heathland east to Surrey with old records from East Sussex and North Essex. *P. scutellaris* is frequent in Wales, known from all vice-counties except Denbighshire, Flint, Monmouth and Montgomeryshire. The beetle is known from Jersey and there are old records from Lundy Island, the Isle of Man, and St. Mary's in the Scillies. Recorded throughout the year with peaks in April and July.

3. *BEROSUS* Leach

This is an unmistakable genus amongst the European Hydrophilidae, with fawn or pale green insects that swim almost continuously. The eyes are bulging. The pronotum is widest at the rear, meeting the elytra to form a more or less continuous body outline. Each elytron has eleven striae, with a short intercalary stria at the front between the first and second striae. The size of the punctures in the striae can usefully be compared with the punctures of the interstices. The upper side has at most a few straggly hairs but the underside has dense, water-repellent pubescence and the legs have long swimming hairs. The sexes can be differentiated by the four-segmented front tarsi of the males, with the two basal segments dilated, other tarsi being five-segmented, females having the tarsal formula 5:5:5. Examination of the male genitalia should not be necessary to make a certain identification, but they can provide additional confirmation. *Berosus* species stridulate when disturbed. The European *Berosus* fauna benefited from the reviews of Schödl (1991, 1993), which resulted in recognition that *B. fulvus* had been misidentified as *B. spinosus*.

Key 6. The species of *Berosus*

1. Conspicuous spine present near the tip of each elytron (viewed from the rear, Fig. 143); last abdominal segment with a smooth outline; head pale except for dark labrum (Fig. 144); pronotum also pale with at most some small, poorly defined smudges; greenish in life, fading to yellow or brown in preserved material; median lobe of aedeagophore distinctive, strongly curved and shaped like a flamingo's head (Fig. 145) ....
.... **4. *Berosus fulvus* Kuwert** (p. 49)

- Spines on elytra absent or inconspicuous; last abdominal segment with indented tip (e.g. as in *B. luridus*, Fig. 146); head dark with metallic patches; pronotum with strong central dark mark or marks (Figs 149 and 155); body colour fawn or brown 2

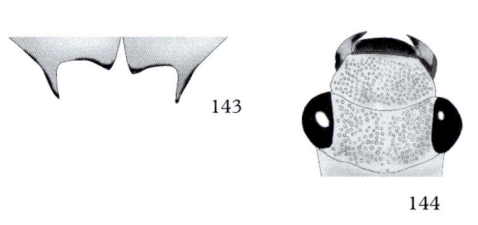

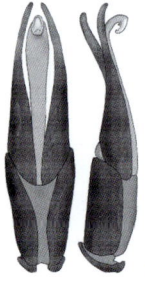

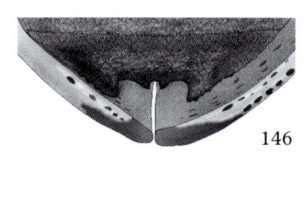

2. Metallic colouration on the pronotum restricted to a patch comprising two dark narrow longitudinal stripes each about the width of a tibia and entirely separated by a pale stripe (Fig. 147), the dark stripes stopping short of the rear edge by more than the width of the pale one; aedeagophore distinctive in that the median lobe articulates with sclerotised and darkened joints at the paramere extremities (Fig. 148); length 4.8 mm or more .... **3. *Berosus signaticollis* (Charpentier)** (p. 48)

- Metallic colouration on the pronotum in a dark, wide patch partly or completely divided in two by a pale, thin stripe (Figs 149 and 155), the patch reaching the rear edge or separated from it by no more than the width of the pale stripe; length less than 4.9 mm 3

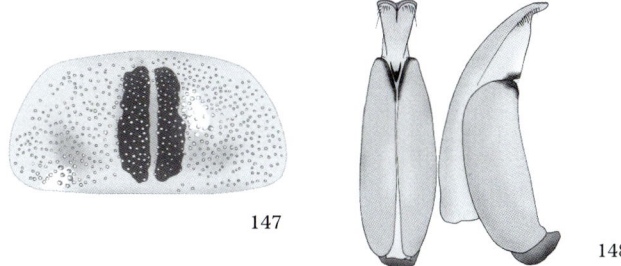

147   148

3. Pronotum with a pale median stripe running at least half the length of the patch, often with a ridge on the surface swollen in its front half (Fig. 149); interstices on the elytra raised into rounded ridges, and with punctures of the interstices almost half as large as those of the striae (Fig. 150); mesosternal ridge with a strong protuberance at its rear end, just in front of the middle coxae (viewed laterally, ventral side to the left, in Fig. 151); aedeagophore similar to that of *affinis*, but larger (1.2-1.4 mm long) (Fig. 154) .... **2. *Berosus luridus* (Linnaeus)** (p. 48)

- Pronotal patch with a pale median stripe extending no more than halfway to the rear (Fig. 155), or the stripe sometimes absent; interstices flat, the punctures of the interstices much smaller than those of the striae (Fig. 156); mesosternal ridge weak without a strong rear projection (Fig. 152); aedeagophore smaller (1.0-1.1 mm long) than that of *luridus* (Fig. 153) .... **1. *Berosus affinis* Brullé** (p. 48)

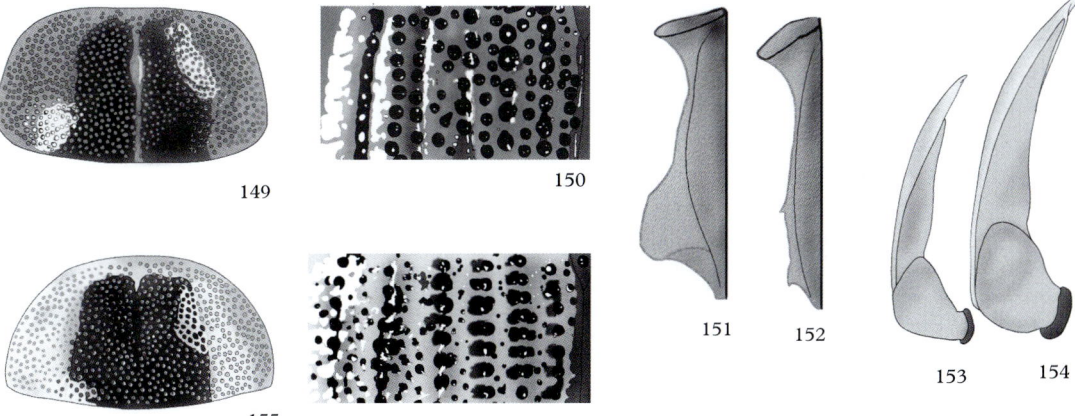

149   150   151   152   153   154

155   156

1. Subgenus *Berosus* Leach

1. *Berosus affinis* **Brullé** Plate 36

Length 3.4-4.8 mm. The difference between the flat interstices of *B. affinis* and the rounded ones of *B. luridus* cannot be depicted convincingly but this is the most accessible feature distinguishing the species. The ground colour of *B. affinis* is paler than in *B. luridus*, this pallor being accentuated by the smaller dark spots surrounding each puncture. The characters of the pronotal patch and the mesosternal ridge are available where it is not possible to compare specimens of the two species side-by-side. The division of the pronotal patch is variable but never as strong as in *B. signaticollis*. *B. affinis* lives in well-vegetated pools and drainage ditches with some exposed substratum. The modern distribution is in low-lying areas south of a line from the Wash to Pembrokeshire, and from there to Dorset, but there are older records to the north, from the Isle of Man, Cheshire, Askham Bog in Mid-west Yorkshire and Potteric Carr in South-west Yorkshire, South Lancashire, Staffordshire, and Derbyshire, and to the south-west in West Cornwall, South and North Devon. *B. affinis* is also known from Jersey. Recorded throughout the year, with peaks in April and September.

2. *Berosus luridus* **(Linnaeus)** Plate 37

Length 4.1-4.8 mm. This species can be confused with the commoner *B. affinis*, the differences being discussed under that species. These species can occur together in fens but *B. luridus* is more typically found over peat. In Ireland *B. luridus* also occurs in base-rich (but nutrient-poor) turloughs, and is confined to the west in Clare, South-east and West Galway, and East Mayo with old records from North Kerry and Roscommon. In Scotland *B. luridus* is formerly known from Easter Ross and Dumfriesshire and was rediscovered in Morayshire in 2010. The English distribution was formerly mainly associated with heathland and old fens north to Askham Bog in Mid-west Yorkshire, but also in the west to Lundy Island. The current distribution is more restricted, reaching to South-east Yorkshire, with other records from 1980 onwards from South and North Somerset, North Wiltshire, Dorset, South and North Hampshire, East Sussex, East and West Kent, Surrey, Cambridgeshire, Hertfordshire, Bedfordshire, West Suffolk, West Norfolk, Northamptonshire, Huntingdonshire, South Lincolnshire, and Leicestershire. The centre of the distribution lies in the old Cambridge Levels and in the Brecks. The only record for Wales is from Glamorgan in the 19th Century. Recorded from February to November, with peaks in April and September.

3. *Berosus signaticollis* **(Charpentier)** Plate 38

Length 4.8-6.1 mm. *B. signaticollis* is often abundant in newly created still water habitats amongst thin vegetation over mud, silt or sand, sometimes in brackish water. In Ireland it is regarded as part of the "moss edge community" (Bilton, 1988) in turloughs and similar habitats in the west in Clare, South-east Galway, Longford and Roscommon. *B. signaticollis* occurs mainly in south and east England north to North-east Yorkshire, with an isolated population on the Lizard in West Cornwall. Welsh records are confined to Glamorgan. It is also known from Jersey. Recorded throughout the year, with peaks in April and October.

2. Subgenus *Enoplurus* Hope

4. *Berosus fulvus* **Kuwert** Plate 39

Length 5.3-6.4 mm, including the elytral spines. The head and pronotum are entirely pale, quite unlike other British and Irish species. Also, *B. fulvus* is distinctly green in life. Its dark labrum or upper lip is also easily seen in the field, the true *B. spinosus* (Steven), with which this species was once confused, having a pale lip. The elytra can be unmarked or bear dark shoulder marks and two rows of marks, the rearmost forming a chevron. This species is confined to shallow water in tidal areas. *B. fulvus* is known only from England, currently in South and North Essex and in West Kent, in the 1980s in East Sussex and East Kent, and earlier in Dorset, the Isle of Wight, South Hampshire, West Sussex, East Suffolk and West Norfolk. Recorded from March to October, with peaks in July and September.

4. *CHAETARTHRIA* Stephens

These are very small and black globular insects, capable of partially rolling up ("conglobation"), and with a row of long hairs covering the first two of the five abdominal sternites and a pair of gelatinous deposits (in Fig. 157 this material has been dislodged from behind the hairs). *C. simillima* was recognised as distinct from *C. seminulum* by Vorst & Cuppen (2003). Separation of these species can only be achieved satisfactorily by examination of the aedeagophore. The sexes cannot be distinguished by external features, though females are a little larger than males. The globular shape of the beetle makes securing them upside down for dissection difficult, Blu-tack® providing a suitable grip under water. It is often best to remove the abdomen and to protrude the aedeagophore by pushing a blunt pin through from the front, thus releasing the structure free from tissue and membranes. The aedeagophore has the stem of the basal piece turned through almost 90° ventrally (ventral side to the right in Fig. 158), such that the structure will not lie flat if placed on its ventral side. The aedeagophore is best viewed dorsal side down and partially cleared in a mountant such as DMHF. The structure is complex with four apparent struts: a pair pointing outwards and derived from sclerotised edges to the parameres and a pair supporting the annulus of the median lobe and pointing inwards. There is also a dorsal strut, termed here a spur, which is quite broad and shorter than the median lobe, therefore falling short of the annulus. The excision on the ventral side of the basal piece appears to be too variable in size to be of use in identification.

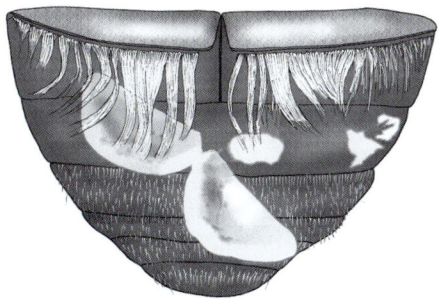

Figure 157. *Chaetarthria,* showing the pair of gelatinous deposits dislodged from behind the row of long hairs covering the first two of the five abdominal.

Figure 158. *Chaetarthria,* the aedeagophore has the stem of the basal piece turned through almost 90° ventrally (ventral side to the right).

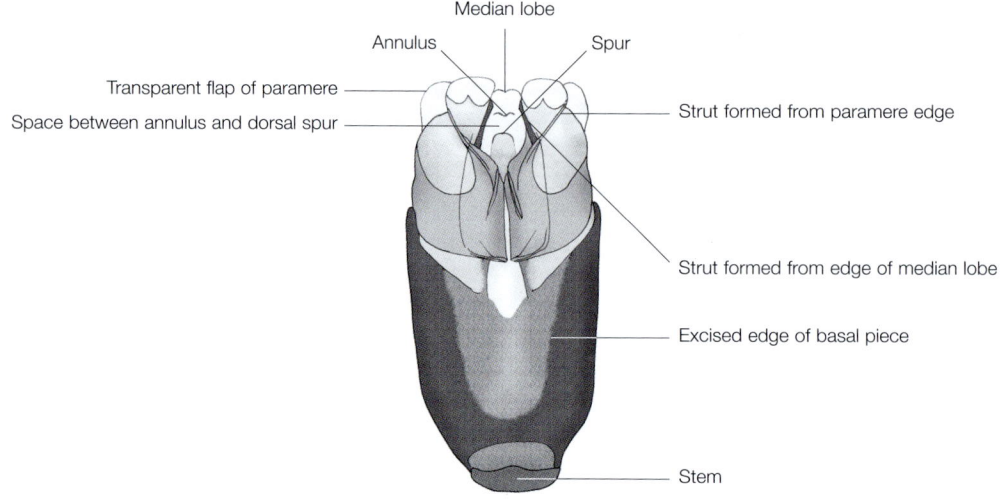

Figure 159. Names of structures of a *Chaetarthria* aedeagophore.

Key 7. The species of *Chaetarthria*

1. Parameres converging and without transparent extensions beyond the struts (Fig. 160); a distinct space between the annulus of the median lobe and the spur ....
.... **1. *Chaetarthria seminulum* (Herbst)** (below)

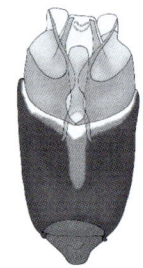

160

- Parameres appearing to diverge owing to transparent flaps extending beyond the struts (Fig. 161); the spur reaching towards the annulus leaving a smaller space ....
.... **2. *Chaetarthria simillima* Vorst & Cuppen** (p. 51)

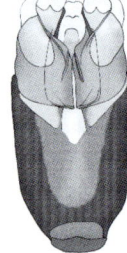

161

1. *Chaetarthria seminulum* (Herbst) Plate 40

Male length 1.3-1.5 mm when curled up, 1.6-2.0 mm when extended. The differences between the aedeagophores of this and *C. simillima* may be as small as are illustrated in Figures 160 and 161, but the gap between the annulus and the spur may be more obvious, as illustrated by Levey (2005). The transparent lobes of the parameres of *simillima* are easily lost if the aedeagophore is allowed to dry out. Externally the only difference between the two species that appears to be consistent is that the head of *C. seminulum* is dull, with dense microreticulation but lacking punctures whereas the head of *C. simillima* is more shining. *C. seminulum* is the rarer of the two species in Britain and Ireland. Although *C. seminulum* can be found amongst moss in rafts of vegetation, also the main habitat of *C. simillima*, it is mainly found in habitats with more exposed mineral substrata, such as silt beds by streams, grazing fen ditches and sites around larger lakes including Loch Lomond, Lough Melvin, Lough Neagh and Lough Owel. *C. seminulum* sensu lato has been recorded across all of Ireland and Britain, from the Scillies to Foula in the Shetlands, but the identification of the individual species has yet to be established over parts of this range. Recorded from April to November, with peaks in May and September.

2. *Chaetarthria simillima* **Vorst & Cuppen** Plate 41

Male length 1.3-1.6 mm when curled up, 1.6-1.9 mm when extended. The habitat is usually amongst moss in seepage, bogs, wet heaths and fens. *C. simillima* is commoner than *C. seminulum* and is widely distributed in Britain and Ireland, but apparently absent from East Anglia. Irish records are from Antrim, Armagh, Clare, West Donegal, Down, West Galway, West Mayo, North and South Kerry, Tyrone, Waterford, and Westmeath. *C. simillima* is recorded in Scotland from Dumfriesshire, Kirkcudbrightshire, Wigtownshire, Renfrewshire, Stirlingshire, Dunbartonshire, Moray, and on Barra, Coll, Lewis, Raasay, Rum, and North Uist. It would appear to be frequent in Wales, with records from Anglesey, Breconshire, Caerfyrddin, Caernarfon, Ceredigion, Glamorgan, and Meirionydd. In England, it is so far recorded from the Scillies, East Cornwall, South and North Devon, North Somerset, Dorset, South and North Hampshire, Surrey, East Sussex, East and West Kent, Middlesex, Berkshire, Warwickshire, North Lincolnshire, South-east, North-east, Mid-west and North-west Yorkshire, and Westmorland. It is also known from Jersey and Guernsey. Recorded in all months except February, with peaks in June and September.

5. *CYMBIODYTA* Bedel

The body shape of this genus is very similar to those of *Enochrus* and *Helochares*. The single European species is small, with its elytra as black as the head and pronotum, characters which should set it apart from the small British and Irish *Enochrus*. This species and *Helochares* have similarly long maxillary palps with the terminal segment a little shorter than the one next to it but *Helochares* have no sutural striae whereas these are found at least on the rear half of the *Cymbiodyta* elytra. *Cymbiodyta* is unusual amongst the Hydrophilidae in having the mid and hind tarsi four-segmented (tarsal formula 5:4:4).

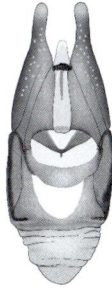

Figure 162. *Cymbiodyta marginellus*, aedeagophore.

1. *Cymbiodyta marginellus* **(Fab.)** Plate 42

Length 3.3-4.3 mm. The name "*marginellus*" may be considered a noun rather than an adjective (the adjective would be *marginellata*) and therefore not required to match the gender of *Cymbiodyta*, the genus to which this was long ago transferred from *Hydrophilus*. *C. marginellus* is easily recognised in the field as being like a small but largely black

Enochrus. The borders of the pronotum and elytra are broadly yellowish brown. However, teneral specimens are uniformly brown and have been mistaken for small *Enochrus*, though the latter would have dark patterns. Examination of the aedeagophore (Fig. 162) will show that this is no *Enochrus* (e.g. Fig. 167) – most characteristic of the genus is wrinkling of what appears to be a sac around the basal piece, and the internal structure is also unusual in having two transverse bridges. *C. marginellus* is frequent in lowland fens in much of England southwards from Westmorland and Teesside, also on Anglesey and along the Welsh coast. In Ireland it is rare north of a line from the Down coast to East Mayo, with an old record from Londonderry and a recent one from Antrim. A cluster of records from the Isle of Man, Kirkcudbrightshire and Wigtownshire, with an old one from Skinburness on the Cumberland coast, link most nearly to the Irish population. Other island records include Alderney, Guernsey, Jersey, and all of the larger Scillies. There are records throughout the year, with peaks in May and September.

6. *ENOCHRUS* Thomson

Enochrus range from 3.1 to 8 mm long, are brown or yellow, have a rounded body outline and are weakly to strongly domed. All have the sutural striae running from the tip of the elytra to at least halfway, other punctures being largely irregular. Three weakly defined rows of punctures are present in the species around *E. quadripunctatus*, and rudimentary punctured rows can be seen on the rear of the elytra of *E. melanocephalus*. Also, the striae may be marked by rows of pigmented spots not coincident with the elytral punctures. The British and Irish species are divided between three subgenera that appear distinct in our fauna and yet are in need of revision, as was noted and planned by the late Stefan Schödl (1998). The singularly black and yellow *Enochrus melanocephalus* stands apart in the type subgenus. The rest of the *Enochrus* species have longer maxillary palps with the basal segment slightly curved outwards (e.g. Fig. 165) and all other hydrophilids have the same segment either oppositely curved or straight. Females are often darker than males, but specimens are most easily sexed by the shape of the claws, distinctly bent inwards in males but more gently curved in females. The aedeagophores are easily extruded and, being flat and thin, can be used for identifying several species. They are best mounted ventral side uppermost so as to be able to see most clearly the position of the annulus of the median lobe in relation to the length of the dorsal spur. The tips of the parameres, which may protrude in preserved material, are sufficiently distinctive without the need for a full dissection of the smaller species in the subgenus *Methydrus*. Unfortunately the aedeagophores are of no use in distinguishing members of the species around *E. quadripunctatus*: it remains likely that *E. fuscipennis*, at one time itself considered to be a form of *quadripunctatus*, is an unresolved species complex.

Key 8. The species of *Enochrus*

1. Head largely black in sharp contrast to the orange or yellow pronotum and elytra (Fig. 163); maxillary palps with terminal segment as long as the next one, basal segment almost straight (Fig. 164); length 4.2-5.1 mm .... Subgenus *Enochrus* Thomson .... **1. *Enochrus melanocephalus* (Olivier)** (p. 57)

- Head either the same pale colour as the rest of the upper surface or head dark with pronotum and elytra brown; maxillary palps with the terminal segment shorter than the next one and the basal segment longest and curved outwards (Fig. 165) 2

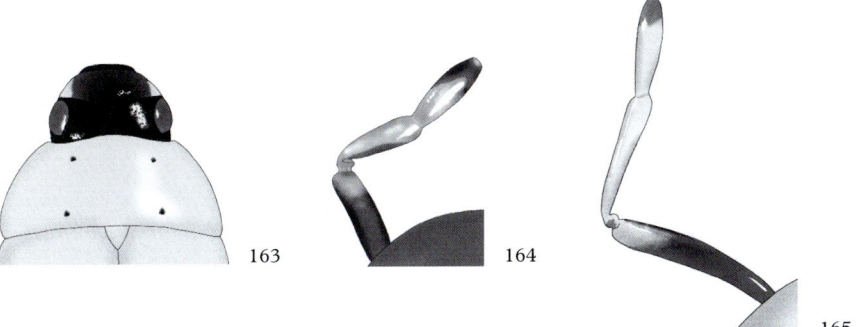

163   164   165

2. Smaller species, 3.0-4.0 mm long; hind margin of last visible abdominal sternite with a semicircular incision filled with strong bristles (Fig. 166) .... Subgenus *Methydrus* Rey
3

- Larger species, 4.2-7.0 mm long; hind margin of last visible abdominal sternite entire .... Subgenus *Lumetus* Zaitzev
5

166

The incision can be smaller than as depicted (Fig. 166) and might only be seen by removing the abdomen and viewing it against a white background.

3. Area in front of eye at most faintly paler than the rest of the head; paramere tips flattened and pointing outwards, and median lobe short with a much longer spur (Fig. 167); length 3.1-3.9 mm .... **8. *Enochrus affinis* (Thunberg)** (p. 60)

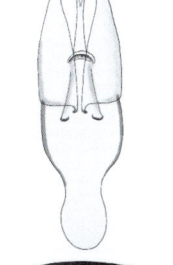

167

- Obvious yellow or brownish mark in front of the eyes (Fig. 168); parameres of the aedeagophore rounded at their tips, median lobe either long or short (e.g. Figs 169 and 171) 4

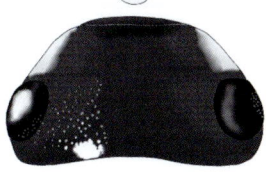

168

4. Median lobe of aedeagophore as long as or longer than the dorsal spur; parameres straight or curved inwards (Fig. 169); dark band filling two-thirds of the space between the suture and the sutural stria (Fig. 170) and most easily seen by lifting an elytron; length 3.4-4.7 mm .... **9. *Enochrus coarctatus* (Gredler)** (p. 60)

- Median lobe much shorter than the dorsal spur; parameres slightly divergent (Fig. 171); darkening confined to the edge of the suture and the sutural stria leaving an area as pale as the rest of the elytra between them (Fig. 172); length 3.2-3.9 mm .... **10. *Enochrus nigritus* Sharp** (p. 60)

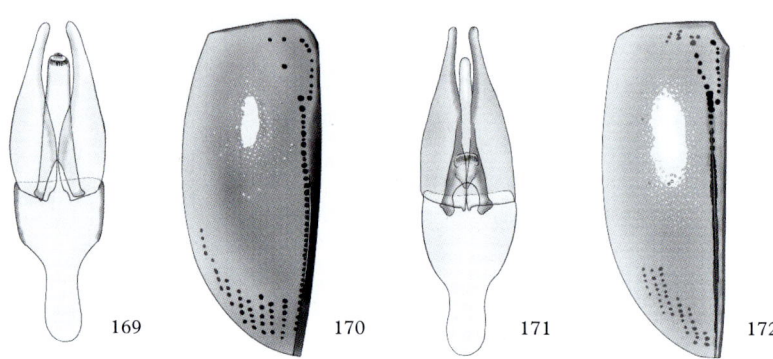

5. Head, pronotum and elytra pale brown or olive, any darker markings being weak; palps all yellow (Fig. 173); median lobe of aedeagophore much shorter than the dorsal spur (Fig. 174); male inner fore claw with outer edge of the basal tooth pointing in the same direction as the claw itself (Fig. 175); length 5.5-6.5 mm .... **2. *Enochrus bicolor* (Fabricius)** (p. 57)

- Head with some strong dark markings; palps with or without dark markings; basal tooth of the male inner fore claw not so well developed that its outer edge points in the same direction as the claw (Figs 178, 179 and 186); median lobe of aedeagophore long or short 6

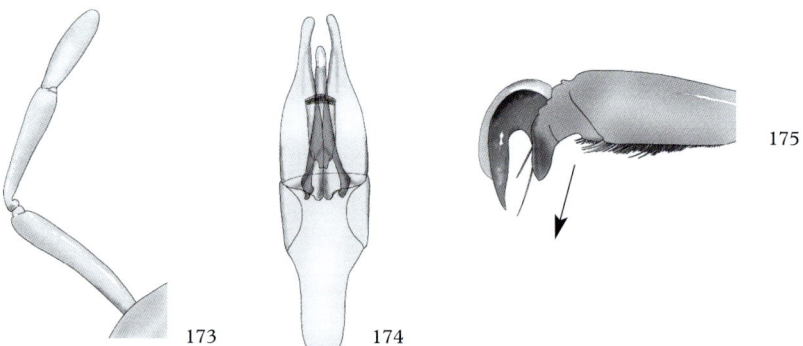

6. Basal segment of the maxillary palps strongly darkened in basal half (Fig. 176); male fore claws Fig. 178; aedeagophore Fig. 177; length 5.4-6.7 mm .... **7. *Enochrus testaceus* (Fabricius)** (p. 59)

- Basal segment of the palps pale 7

   Make sure that the palp is completely unfolded to view the basal segment. Note that this is not the true basal segment, which is very small and not visible from above.

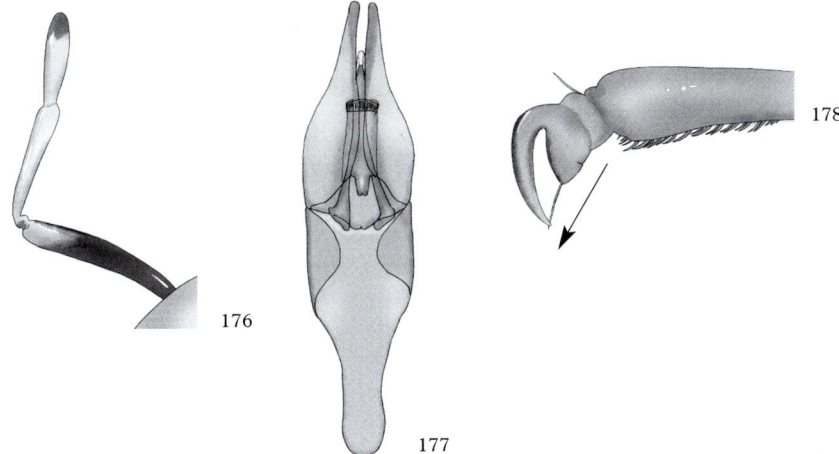

7. At least the front half of the elytra without a trace of rows of punctures; male inner front tarsal claws each with a weak basal tooth (Fig. 179); aedeagophore slender with dorsal spur protruding well beyond median lobe (Fig. 180); male with the tip of the maxillary palp pale (Fig. 181), tip usually darkened in female (Fig. 182); length 4.7-5.5 mm .... **5. *Enochrus ochropterus* (Marsham)** (p. 58)

- Elytra with two or three longitudinal irregularly spaced rows of punctures (Fig. 183 - view at × 20; light from the side, on a **dry** beetle); male inner front tarsal claws with a large tooth at the base (Fig. 186); aedeagophore with median lobe almost as long as the dorsal spur (Fig. 187); tips of the maxillary palps dark or pale 8

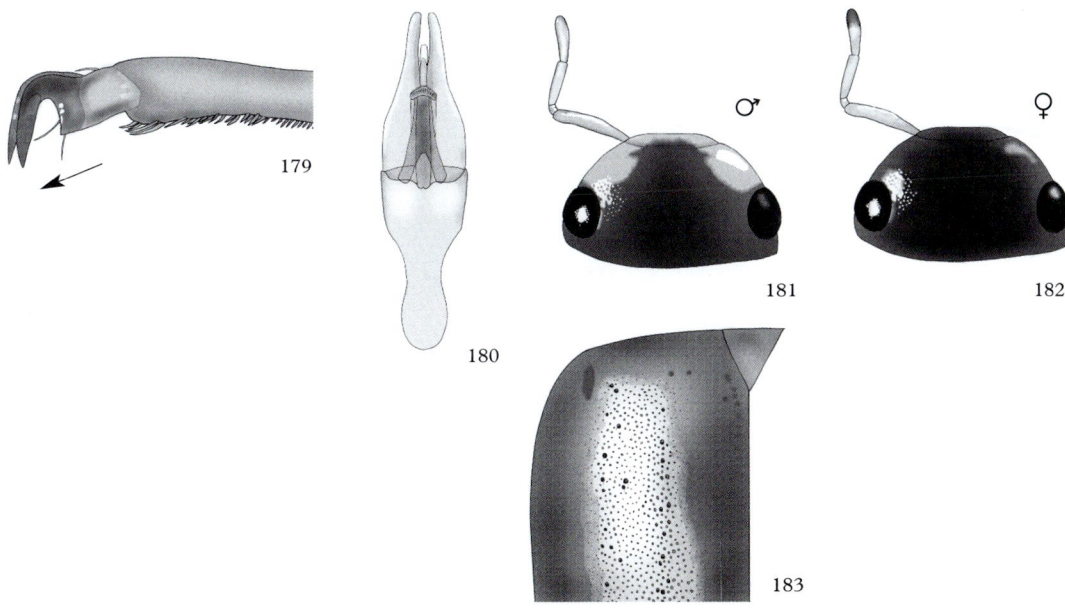

8. Palps pale, sometimes slightly darkened near tips (Figs 184 and 185); aedeagophore Fig. 187; length 4.8-5.9 mm .... **4. *Enochrus halophilus* (Bedel)** (p. 58)

- Tip of the last segment of the palp dark in both sexes 9

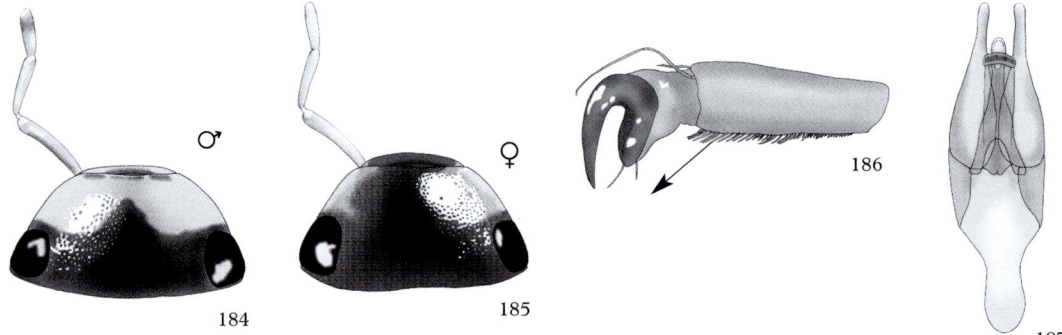

9. Pronotum with a central black spot and four smaller spots on the corners of an imaginary trapezium around it (Fig. 188); labrum orange in male (Fig. 189) and black in female (Fig. 190); length 4.7-5.7 mm .... **6. *Enochrus quadripunctatus* (Herbst)** (p. 59)

- Central black spot of pronotum large, at the very least encroaching on the smaller spots (Fig. 191); colouring of head mostly or entirely black with very limited sexual variation, last segments of palps darker; length 4.2-6.2 mm ....
.... **3. *Enochrus fuscipennis* (Thomson)** (p. 58)

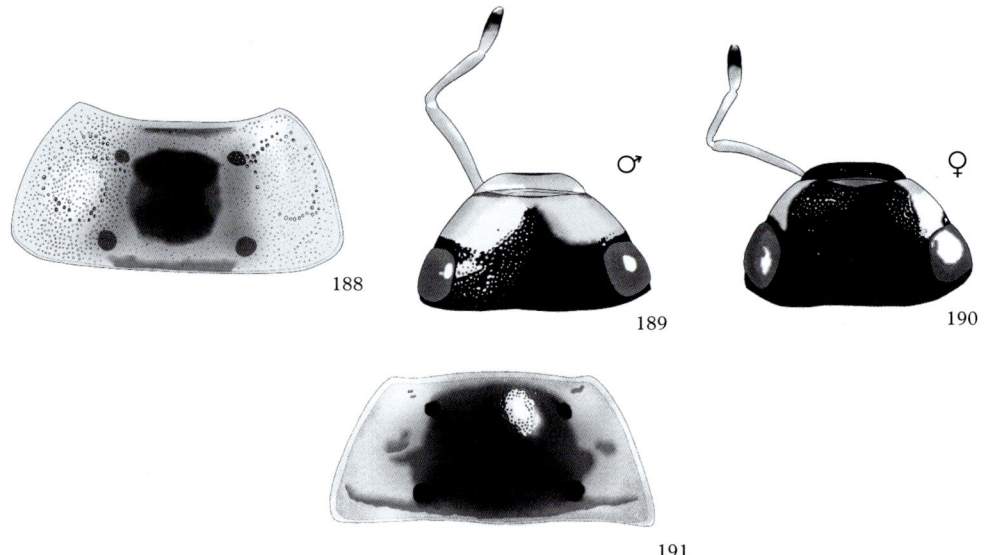

Subgenus *Enochrus* Thomson

The type subgenus has one distinctively coloured species with comparatively short maxillary palps.

1. *Enochrus melanocephalus* (Olivier) Plate 43

Length 4.2-5.1 mm. This species is easily recognised, even in the field, because of its bi-coloured appearance, with the orange pronotum and elytra contrasting strongly with the black head, underside and legs. The habitat is lowland, thinly vegetated standing waters, typically disturbed or of recent origin and often polluted or brackish, with high conductivity. *E. melanocephalus* is frequent in the English Midlands, in the Thames Valley and in grazing fens around the coast. It has recently expanded through the north-east of England but is still rare in the north-west, recorded in West Lancashire in 2007 – and from the Isle of Man in 2009. In Scotland *E. melanocephalus* was first reported, in Kirkcudbrightshire, in 2004. The Welsh records are mainly extensions of the inland English distribution, with outlying records from Anglesey and Glamorgan. In Ireland there are records along the south coast in East and Mid Cork, Waterford, and Wexford, and scattered records further north, in County Down, Kilkenny and Longford. This beetle is known from Bryher and Tresco in the Scillies, and from Guernsey and Jersey. Recorded throughout the year, peaking in May and September.

Subgenus *Lumetus* Zaitzev

Lumetus is effectively the catch-all subgenus of medium-sized species with long maxillary palps and without the incision on the last abdominal sternite. The subgenus might also be characterised by the "pronotal ellipses", as shown in Fig. 188, weakly defined rings of coarser punctures on the side of the pronotum. The latter can be very difficult to see and are of no use in determining individual species. All of the species here show considerable variation in colouring and yet, perversely, colour provides some of the best characters other than the aedeagophores, which divide broadly into the *E. quadripunctatus* complex, with a long median lobe, and the rest, in which the dorsal spur is more obvious.

2. *Enochrus bicolor* (Fabricius) Plate 44

Length 6.5-7.5 mm. In the field *E. bicolor* can be recognised by the general pallor of its dorsal side, though it can occasionally have smudge-like markings on the centre and rear of the head, on the pronotum and on the elytral shoulders. The maxillary palps are long and pale (Fig. 173), but not as long in *Helochares lividus*, a species with which *E. bicolor* is sometimes confused. The aedeagophore (Fig. 174) is not especially distinctive but it should at least distinguish it from the other brackish water species, *E. halophilus*. *E. bicolor* is only common in brackish water though its propensity for flight results in individual specimens being found in other habitats. Thus it is largely confined to coastal ponds and slow-flowing ditches, but can also be found inland in Cheshire and Worcestershire where pools receive saline seepage. Butler and Popham (1958) noted its association with pools with water at a concentration of more than 50% seawater on the Spurn peninsula. Greenwood and Wood (2003) noted occurrence in Essex over a salinity range of 5 to 63 parts per thousand (ppt) (*cf.* 35 ppt for seawater), the maximum range observed over the four-year study. *E. bicolor* can be found around the coast of Ireland south from Galway and County Down, the latter slightly further north than the limit of this species in England on Teesside in County Durham. It reaches north on the west coast only as far as Anglesey. Arribas *et al.* (2012) have drawn attention to the great dispersal ability of *E. bicolor*, making it all the more surprising that this species has not reached the Solway. *E. bicolor* is commonest from the Humber to the Solent. It is known from Guernsey and the Scillies. Recorded in all months except December, with peaks in June and August.

3. *Enochrus fuscipennis* (Thomson) Plate 45

Length 4.2-6.2 mm. This taxon is variable in the extent of darkening with mitochondrial DNA and karyotypes suggesting that it in fact a species complex in northern Europe. The extent to which the complex is represented in Britain and Ireland is uncertain. Irish specimens tend to be more brightly marked and Highland specimens darker and more diffusely marked than in those from the south. *E. fuscipennis* is confined to acid water, usually on peat or amongst *Sphagnum*, and can occur from sea level to over 600 metres elevation. The distribution is mainly western, being frequent in Ireland and ranging from the Lizard in Cornwall to Unst in the Shetlands, including all mountainous areas and most islands. There is also a band of records running from Dorset to East Norfolk, and *E. fuscipennis* is frequent on the North York Moors, though rare further north on the English and Scottish east coast. There are major gaps in the distribution in the south of Ireland away from Kerry, in the Weald, in the English Midlands and in the Scottish industrial belt. Recorded in all months except December, peaking in June.

4. *Enochrus halophilus* (Bedel) Plate 46

Length 4.8-5.9 mm. This is a difficult species to determine with certainty as the colour characters that one must use are variable, and have been for the most part poorly defined in the past. The male head is typically yellow with the area between the eyes black and a black triangle in front of it not reaching the labrum, which can be darkened along its internal edge and in the middle (Fig. 184). The maxillary palp may be entirely yellow or may have the middle of the last segment and the outer side of the basal segment slightly tinged. The female head is extensively darkened leaving large yellow marks in front of the eyes (Fig. 185). The palp is rarely in any way darkened in female *E. halophilus*, running contrary to the rule that *Enochrus* females are darker than males. More consistent features are that the centre of the pronotum is darkened but never black, and that the scutellum and area around it are as pale as, or more typically paler than, the rest of the elytra. *E. halophilus* is confined to brackish water not just on the coast but also where there is brackish seepage inland. The northernmost record is on Teesside in County Durham, and *E. halophilus* is recorded almost continuously from the Humber valley in North-east, South-east and Mid-west Yorkshire, North and South Lincolnshire, the hinterland of the Wash in Cambridgeshire, Huntingdonshire, West Norfolk and Northamptonshire, also around the coastal vice-counties to the Isle of Wight and Dorset. Other coastal records are for Anglesey, Caernarfon, Caerfyrddin, Pembrokeshire, Glamorgan, North Somerset, and West Cornwall including St. Agnes, St. Mary's and Tresco in the Scillies, and also on Jersey. Inland the species is known from Cheshire and Nottinghamshire. The northernmost record for Ireland is from Clare, others being from North Kerry, Mid Cork and Wexford. Recorded in all months except February, with peaks in May and July.

5. *Enochrus ochropterus* (Marsham) Plate 47

Length 4.7-5.5 mm. A nondescript species, possibly the most distinctive feature being the way in which the sexes differ between each other in colouring. The males have at least the sides of the head pale, the pronotum with its central patch often brown rather than black, and the scutellum the same brown colour as the elytra. Females have the head (including the labrum) black though there are sometimes paler patches in front of the eyes, and the scutellum is mostly black. The tips of the palps are typically blackened. It is the most likely species to be confused with *E. fuscipennis*, which is dark in both sexes, and has two or three rows of coarser punctures on the elytra. The body shapes of the two species differ, *E. ochropterus* being broader and more highly domed. Although this difference is difficult to define it provides, with experience, a good field character. Males of *E. ochropterus* and *E. fuscipennis* can be separated on the length of the median lobe of the aedeagophore and on the development of the basal tooth of the inner fore claw. *E. ochropterus* is typical of mesotrophic fen, usually amongst *Sphagnum* and other mosses with plenty of plant litter. The habitat may be found in lakes and mires, particularly where base-flushing occurs over peat, such as in abandoned peat cuttings or in groundwater-fed bogs. *E. ochropterus* is widespread across mainland Ireland and in Britain north to the Black Isle. On the west coast the northernmost record is from Tiree other islands reliably recorded being Anglesey, Guernsey, Jersey, Little Cumbrae, the Isle of Man, Skokholm, Tresco, and the Isle of Wight. Recorded throughout year, peaking in April and August.

6. *Enochrus quadripunctatus* (Herbst) Plate 48

Length 4.7-5.7 mm. Despite being part of an as yet unresolved complex of species *E. quadripunctatus* is distinctive even in the field, with the dark markings of the head and pronotum contrasting with the rest of the upper side which is yellowish brown or yellow. The central pronotal spot never quite engulfs the smaller spots and the palps have sharply defined black tips in both sexes. The aedeagophore and the male inner fore claw do not distinguish *E. quadripunctatus* from *E. fuscipennis* and *E. halophilus*. A feature that can be seen in Figs 189 and 190, the exposure of a membrane between the labrum and the V-shaped front margin of the clypeus, has been used to distinguish this species from *E. fuscipennis* on the basis that its clypeal edge is straighter. However, there appears to be too much variation in the edge's shape to make this reliable. *E. quadripunctatus* is a mobile species, readily taking to flight, and occurring in lowland, base-rich stagnant water with some exposed mineral substratum, also in mesotrophic fens. It appears to have expanded its range recently. The distribution is mainly eastern in Britain, in most vice-counties from East Sussex to two Scottish sites in Fife and East Lothian, with the beetle most frequent around London and on the Brecks. Records in the west are for Bryher and St. Mary's in the Scillies, the Lizard, South Devon, South and North Somerset, and Caerfyrddin. There are inland records from Oxfordshire, Warwickshire and Nottinghamshire, and *E. quadripunctatus* has also been recorded from Jersey. *E. quadripunctatus* is not known from Ireland. Recorded from March to October, peaking in April and August.

7. *Enochrus testaceus* (Fabricius) Plate 49

Length 5.4-6.7 mm. As the largest *Enochrus* in freshwater in Britain and Ireland, *E. testaceus* should pose no identification problems once the darkened basal segment of the maxillary palp is seen. In addition to their strongly bent claws, males can be distinguished from females by the labrum being entirely yellow whereas it is centrally darkened in females. *E. testaceus* is found in fens and richly vegetated ponds, lakes and ditches in lowland areas. In Ireland most of the records are for the karstic area running from County Down to Limerick, with another cluster of records in the south-east, and probable absence from the north coast. The British distribution is dominated by its frequency in the English lowlands with outliers in Teesside and the south of the Lake District. However, the most extreme outliers are at Aberlady Bay, East Lothian, in several sites in Kirkcudbrightshire, and an old record needing confirmation from West Perthshire. Welsh records are mainly from Anglesey and the English border counties, plus one site in Ceredigion and old records from Glamorgan. *E. testaceus* is also known from Guernsey and Jersey. Recorded throughout the year, peaking in May and August.

Subgenus *Methydrus* Rey

This subgenus is in northern Europe composed of small species with a small semicircular incision in the middle of the rear edge of the last visible abdominal sternite (Fig. 166).

8. *Enochrus affinis* (Thunberg) Plate 50

Length 3.1-3.9 mm. *E. affinis* can be distinguished from the other small *Enochrus* by its dark features, such as the darkened terminal segments of the palps and the lack of pale marks in front of the eyes. However, the area between the suture and the neighbouring striae is not usually filled with pigment as it is in *E. coarctatus*. The distinctively bent tips to the parameres may protrude enough to avoid the need for dissection. *E. affinis* is mainly associated with floating or flooded *Sphagnum*, and it can be abundant where the moss sward has been damaged: it can coexist with *E. coarctatus* but is more often found in association with *E. fuscipennis* or *E. ochropterus*. In the north of its range it is found in a wider range of still water habitats, including temporary grassy pools. The distribution is predominantly western, it being frequent across much of Ireland, Wales, mainland Scotland, and the Lake District. *E. affinis* is also well recorded from Yorkshire, more so on Thorne and Hatfield Moors and other low-lying peatlands than in the Pennines. In the south *E. affinis* is known from the band of heathland stretching towards London from Dorset, also from a few sites in East Cornwall, South Devon, and Somerset, though some old records may refer to *E. nigritus*. The island distribution is patchy – Anglesey, Arran, Clare Island, Coll, Colonsay, Eigg, Islay, Jura, Mull, Ornsay, Raasay, Rum, Skye, and the Uists. Recorded throughout year, peaking in June and August.

9. *Enochrus coarctatus* (Gredler) Plate 51

Length 3.4-4.7 mm. In the field the dark brown sutural stripe distinguishes this species from other small *Enochrus*. The aedeagophore's long median lobe provides the best confirmatory character. The habitat is in eutrophic or mesotrophic fen, including well vegetated edges of ponds and sluggish ditches. *E. coarctatus* is frequent over much of lowland Britain and Ireland, but is almost absent from the Scottish Highlands, reaching to South Aberdeenshire on the east coast and with isolated records from Barra, Tiree, Lismore and the adjacent Argyll mainland in the west. It is also common on Islay and found on Bute, Cumbrae and Little Cumbrae in the Clyde Isles, but not on Arran. Other islands with records include Anglesey, Jersey, Man, St. Mary's in the Scillies, and Sherkin Island. Recorded throughout the year, peaking in May.

10. *Enochrus nigritus* Sharp Plate 52

Length 3.2-3.9 mm. *E. nigritus* occurs in its dark form on mountains in Spain and Portugal, the paler form, once described as a distinct species, *E. isotae* Hebauer, being found in lowlands from around the Mediterranean and through western France to Britain. The British form is most likely to be confused with *E. coarctatus*, but the suture is not boldly striped, darkening being confined to the elytral edge and the sutural stria. The parameres are finger-like as in *E. coarctatus* but the median lobe is short as in *E. affinis*. This pale form of *E. nigritus* has pale marks in front of the eyes as in *E. coarctatus* but the outer two-thirds of the terminal segment of the maxillary palp are black, whereas this segment has the tip slightly paler than the brown middle third in *E. coarctatus*. The habitat is mesotrophic and base-rich fen, most often in relict sites and in association with aquatic mosses. In Wales it is known from Anglesey, being otherwise English, from South and

North Hampshire, East Sussex, Berkshire, Oxfordshire, West Suffolk, East and West Norfolk, Cambridgeshire, Huntingdonshire, Herefordshire, and Cheshire. Recorded throughout the year, peaking in April, July and September.

7. *HELOCHARES* Mulsant

The three species found in Britain, one of which occurs in Ireland, belong to the subgenus *Helochares* Mulsant. This treatment follows Hansen (1982), who was first to recognise that there were three species in northern Europe. The most obvious feature of these moderately sized (4-6 mm long) hydrophilids is that the maxillary palps are very long (Plates 53-55). The northern European species of *Helochares* share with *Laccobius* the absence of sutural striae. A particularly obvious field character is that the females carry their egg sacs attached to the underside of the abdomen, a behaviour seen elsewhere in British Hydrophiloidea only in *Spercheus emarginatus*, a species extinct in Britain.

The dorsal puncturing of *Helochares* provides important identification characters. There are two types of punctures, the abundant and evenly scattered ones of the surface and some larger ones arranged in weakly defined semicircles on the sides of the pronotum and in several lines on the elytra (Fig. 192 is a side view of *H. lividus* showing only these larger punctures). The larger punctures are setiferous, bearing long straggly hairs (e.g. Fig. 198) that lie flat on the surface and can be seen at a magnification of × 20 or more. These setiferous punctures are infrequent in *H. obscurus* and cannot be seen easily amongst the closely packed punctures lacking obvious setae.

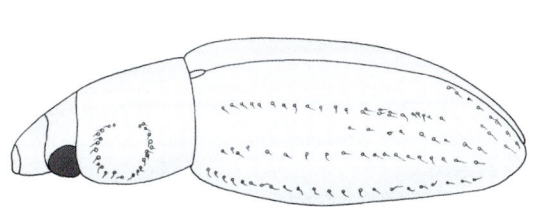

Figure 192. *Helochares lividus*, side view showing larger punctures on pronotum and elytra.

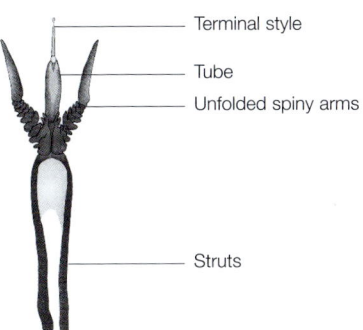

Figure 193. *Helochares*, median lobe.

Males are slightly smaller and narrower than females and have enlarged claws each with a basal tooth. The aedeagophore is abundantly distinct from that of other hydrophilids, with a very short basal piece, and the parameres forming a tunic (see Fig. 196) around the median lobe, which has spiny arms and a tube bearing a terminal style. The struts of the median lobe are prominent in this genus, providing the attachment points for the muscles that evert the lobe's spines, as was demonstrated by Balfour-Browne (1958). It would be possible to differentiate species using the median lobe when extended but its fine structure is obscured when at rest and under the normal conditions of preservation. The median lobe is shown dissected here (Fig. 193) in *H. punctatus* in order to identify the various parts: however, dissection of the parts of the aedeagophore is not recommended. The relative positions of the parameres and the median lobe at rest can be used to differentiate species, but differences in the overall size of the aedeagophore, as has been suggested in the past, are too marginal to be of use in identification.

Key 9. The species of *Helochares*

1. Upper side of body pale, usually yellow or greenish in life, with the labrum as pale as the rest of the head, and the extremities of the maxillary palps slightly darkened (Fig. 195); aedeagophore with no structures of the median lobe reaching the paramere tips (Fig. 196); main elytral punctures very fine, the distance between the punctures much larger than their diameter, which is generally half or less that of the pores of the setiferous punctures (Fig. 194), which are present in semicircles on the pronotum and in partial rows on the elytra (Fig. 192); length 4.5-5.8 mm .... **1. *Helochares lividus* (Forster)** (opposite)

- Upper side of body darker with the labrum dark and a third or more of the tips of the maxillary palps distinctly darkened (Fig. 197); parts of the median lobe reaching as far as the paramere tips (Figs 199 and 201); elytral punctures spaced at no more than their diameter's width apart (Figs 198 and 200); setiferous punctures more difficult to see amongst the coarser punctures, sometimes not visible at all 2

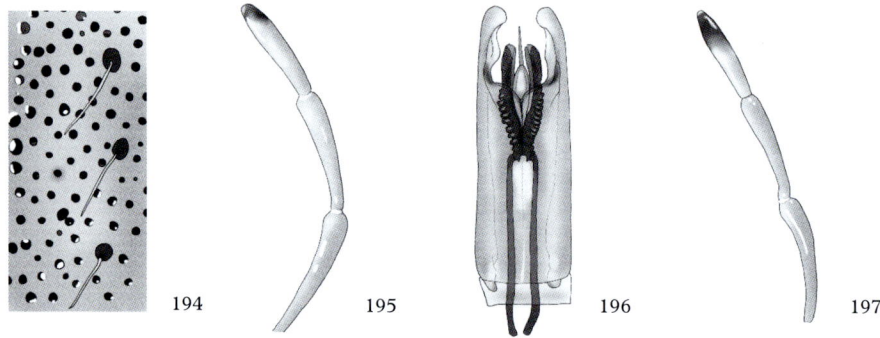

194   195   196   197

2. Head black or at least with the midline blackened; elytra with setiferous punctures in rows, the rest of the large punctures spaced at about a diameter apart from each other ("stand alone" as in Fig. 198); most of the tube of the median lobe protruding beyond the paramere "tunic" (Fig. 199); length 4.5-6.0 mm .... **3. *Helochares punctatus* Sharp** (opposite)

- Head usually less dark than the labrum; few or no setiferous punctures on the pronotum and elytra, the other punctures of the upper surface so dense that nearly all distort the shapes of their neighbours (Fig. 200), with the gaps between them less than the diameter of the punctures; about half of the tube of the median lobe protruding beyond the paramere tunic (Fig. 201); length 4.4-5.8 mm .... **2. *Helochares obscurus* (Müller)** (opposite)

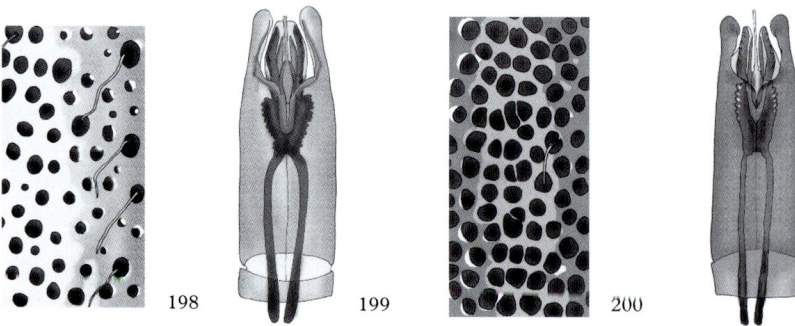

198   199   200   201

1. *Helochares lividus* (Forster) Plate 53

Length 4.5-5.8 mm. *H. lividus* is the easiest of the three species to recognise, being pale with fine punctures on a distinctly shiny surface. This species has sometimes been confused with *Enochrus bicolor*, the palest of the *Enochrus*, about the same size and with long maxillary palps, though not as long as in *H. lividus*. *H. lividus* lives in vegetated lowland freshwaters, often in areas with a brackish influence. The distribution is compact in Wales and England, north to the Fylde in West Lancashire and Pickering Vale in North-east Yorkshire, with a recent incursion into County Durham. *H. lividus* is frequent in the Scillies and the Channel Isles. Recorded throughout the year, with peaks in May and September.

2. *Helochares obscurus* (Müller) Plate 54

Length 4.4-5.8 mm. This is one of the most difficult hydrophilid species to identify with certainty, mainly because the species with which it was once confused, *H. punctatus*, is variable in the strength of its punctures and colouring. The final decision will usually be based on the virtual absence of rows of setiferous punctures on the elytra. The tube of the median lobe is smaller than that of *H. punctatus* and at rest is seated lower between the parameres than in *H. punctatus*. *H. obscurus* is found in relict mesotrophic fen, and is known from West and East Norfolk, Huntingdonshire, and Herefordshire, with a 19th Century record from Whittlesey Mere in Cambridgeshire. Recorded in all months except November, with peaks in May and September.

3. *Helochares punctatus* Sharp Plate 55

Length 4.5-6.0 mm. *H. punctatus* is easy enough to identify if dark and found in its main habitat, on moist peat in wet heath, in bogs and in acid pools. Paler specimens have been misidentified as *H. obscurus*. *H. punctatus* occurs in the southern part of Ireland north to Achill Island in West Mayo. The main part of the British distribution is in two wide south-west to north-east swathes from East Cornwall to the Broads and from Wales, where this species is frequent, to the North York Moors. The intervening land has little suitable habitat but even where this occurs, as in the Somerset Levels, this species appears to have died out. Outlying areas of occupancy are in the south of the Lake District, low-lying mosses around Carlisle, the Solway coast along to Wigtownshire, Colonsay, Jura and Kintyre. *H. punctatus* is also known from Guernsey and Jersey, Lundy and the Isle of Man. Recorded throughout the year, with peaks in April and August/September.

8. *HYDROBIUS* Leach

In Britain and Ireland *Hydrobius* is represented by what is almost certainly a complex of medium-sized (6-8 mm long), brown or black convex species, with each elytron bearing 10 striae, strongly grooved in the rear half but fading to weak series of punctures at the front. The main water beetle genus with which this complex might be confused is *Limnoxenus*, *L. niger* (Gmelin) being larger, usually exceeding 8 mm in length, with the elytral rows of punctures not in grooves except for those either side of the suture in the posterior half. The metasternum is not ridged as in *Limnoxenus*. Apart from *fuscipes* s. lat. the other European species are *H. arcticus* Kuwert, confined to the Fennoscandinavian tundra, and *H. convexus* Brullé, a much larger species (9.5-12.0 mm long) of the extreme edges of ponds and ditches, which reaches north to Loire Atlantique. The front tarsal claws are more strongly curved in males than in females but sexing is usually made easier by sighting the tips of the female's cerci. Given the likelihood that more than one species is involved it is important to retain vouchers of this species and it may ultimately prove desirable to dissect out the aedeagophore.

1. *Hydrobius fuscipes* (Linnaeus) Plate 56

Length 6.4-8.7 mm. *H. fuscipes* is easily recognised in the field, other dark convex water beetles being either dytiscids or hydrophilids, the latter being either much smaller (*Coelostoma orbiculare*) or larger (*Limnoxenus niger*). Mitochondrial DNA indicates that *H. fuscipes* as currently devised comprises more than one species (Johannes Bergsten, pers. comm.). The type form and var. *subrotundus* Stephens (a senior synonym of *H. picicrus* Thomson and *subrotundatus*, a misspelling of Stephens's name for this form) share the same arrangement of larger, hair-bearing punctures. These are on the 2nd, 4th, 6th and 8th interstices: they differ in that the type form has paler legs and is less hunched in posture. Another taxon is var. *rottenbergi* Gerhardt, which is similar to the type form in colouring but has the hair-bearing punctures lying in or adjacent to the 3rd, 5th, 7th and 9th striae (not counting the short intercalary row between the 1st and 2nd striae) (Fig. 203), i.e. not on the interstices as in the other forms (Fig. 202). Two names, var. *chalconotus* Stephens and var. *aeneus* Solier, appear to refer to the same form with a strongly metallic (usually green or blue, but also bronze) colouring. Both the type form and *subrotundus* can be found widely, the former to be found in almost any lowland wetland, whereas the latter is confined to acid waters, typically in *Sphagnum* pools. Specimens with the hair-bearing punctures of the elytra described above for *rottenbergi* have been detected from South Hampshire, West and East Sussex, East Kent, Cambridgeshire, Herefordshire, and Westmorland. Hansen (1987) notes the occurrence of *rottenbergi* in brackish water in the south of Scandinavia, and specimens apparently belonging to this form are known from around the Mediterranean. Despite its ubiquity, this complex is often identified by community analysis as an indicator of temporary water. Members of the complex are common throughout Britain and Ireland north to the Orkneys, being found throughout the year with a peak in May.

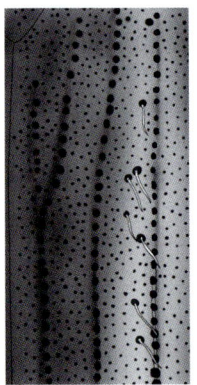

202

203

9. *LIMNOXENUS* Motschulsky

The British representative is large (8 mm or more long), black and convex, and, unlike *Hydrobius*, capable of swimming. It should be possible to see in the field, with or without the aid of a × 10 hand lens, that the thoracic sternites are strongly ridged centrally with the front half of the metasternum developed into a keel extending towards the mesosternal keel (Fig. 38). The best dorsal character is that the 10 rows of punctures on each elytron do not run in grooves, except for the rear half of the sutural striae (Fig. 37). Live individuals often stridulate when handled. The only other European species is the Iberian *L. olmoi* Hernando & Fresneda.

1. *Limnoxenus niger* (Gmelin) Plate 57

Length 8.0-9.8 mm. This species might only be confused with larger specimens of the *Hydrobius fuscipes* complex. *L. niger* is almost entirely coastal in England and South Wales though it is not confined to brackish water, occurring in well vegetated ponds and drainage ditches. The northernmost record is for the coast of West Norfolk and it is frequent in the Broads, on the Thames Marshes, on the Kent and Sussex grazing fens, the Somerset Levels and the adjacent parts of Monmouthshire. There are old records inland from the Cambridgeshire fens and it is still present at Woodwalton Fen, Huntingdonshire. It is common on Guernsey. Recorded from February to November, with strong peaks in April and September.

10. HYDROCHARA Berthold

European *Hydrochara* species are second largest in size to *Hydrophilus* among the Hydrophilidae. *Hydrochara* also can be characterised by the spine of the thoracic keel not reaching the second visible abdominal sternite as in *Hydrophilus*. The only British species is *H. caraboides*. The other European species are *H. flavipes* (Steven) and *H. dichroma* (Fairmaire). The former, which is easily recognised by its pale legs, comes nearest to Britain in southern France, whereas the latter is nearest in Hungary and Greece.

1. *Hydrochara caraboides* (Linnaeus) The Lesser Silver Water Beetle Plate 58

Length 13.0-18.5 mm. *H. caraboides* is a strong swimming species living in permanent, well vegetated ponds and drainage ditches. At up to 11 mm *Hydrobius convexus*, a species reaching north to Loire-Atlantique, might be confused with the *Hydrochara*, but it lacks the pointed keel. There is a 19th Century record of *H. caraboides* for Glamorgan and from Ceredigion in 1922, but all modern Welsh records are for Denbighshire adjacent to the Cheshire Plain, where this species has been known since 1990. Until then *H. caraboides* was only reported regularly for many years from the Somerset Levels. Older records are from West Sussex, Surrey, Middlesex, South and North Essex, Cambridgeshire, Huntingdonshire, Derbyshire, Mid-west Yorkshire, and South Lancashire, and subfossil records for Worcestershire and North Lincolnshire. Adults of *H. caraboides* have been recorded from March to November. The Cheshire Plain population peaks in July whereas that in Somerset peaks in April and May. Larvae generally occur from May to August.

11. HYDROPHILUS Linnaeus

These are the largest European water beetles. The British representative is *H. piceus*. Given the capacity of *Hydrophilus* to fly, it is worth bearing in mind that vagrants of the more eastern species, *H. aterrimus* (Eschscholtz) and the more southern one, *H. pistaceus* (Castelnau) might easily be overlooked as differences between the three European species are small. The median ridges on the abdominal sternites are sharp in *H. piceus* and *H. pistaceus* but weak in *H. aterrimus*. *H. piceus* has a spine projecting 0.15 mm at the tip of each elytron (Fig. 204); sometimes these spines are broken but their stubs can still be seen.

H. aterrimus has blunt spines and *H. pistaceus* none at all. The fore claws of male *Hydrophilus* have the last segments drawn out into plates: that of *piceus* is rounded (Fig. 205) as in *aterrimus*, whereas that of *pistaceus* is more pointed.

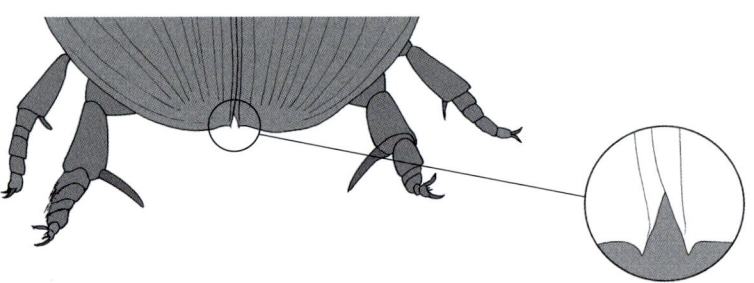
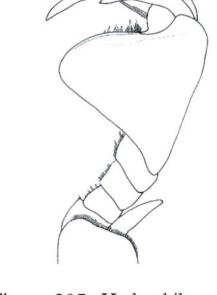

Figure 204. *Hydrophilus piceus*, showing spine projecting 0.15 mm at the tip of each elytron.

Figure 205. *Hydrophilus piceus*, fore claw of male.

1. *Hydrophilus piceus* Linnaeus The Great Silver Water Beetle Plate 59

Length 34-48 mm. Spurious records of *H. piceus* are occasionally generated by non-coleopterists simply because of the publicity associated with its size, but this species can scarcely fail to be recorded correctly otherwise. The faint possibility of other *Hydrophilus* species is discussed above. *H. piceus* breeds in permanent, richly vegetated ponds and in dykes on grazing fen but individuals have been found, presumably after flight, on greenhouse roofs, beneath street lights, at light traps and even on an oil rig in the North Sea. Modern records are from England in both vice-counties of Somerset, Kent, Sussex, Essex and Norfolk, also in East Suffolk. Older records are from Dorset, the Isle of Wight, South Hampshire, Surrey, Middlesex, Cambridgeshire, Huntingdonshire and Leicestershire, a record from medieval remains extending the range to South-west Yorkshire, and a 19th Century record indicating that it once reached Newcastle-upon-Tyne. There are old records for Guernsey and Jersey. In Wales *H. piceus* is found on the Monmouthshire Levels, with a 19th Century record from Glamorgan. There is a doubtful record from Ireland in Down, recounted by Balfour-Browne (1962), who noted that he had reared specimens in his garden in the same county. The life-cycle has frequently been researched. Adults have been recorded in all months except January, peaking in June and August with a distinct dip in July. Larvae have been reported from April to September.

12. *LACCOBIUS* Erichson

These are small insects, none scarcely more than 4 mm long. They are typically teardrop-shaped and convex, with a bow-legged appearance given by the curved hind tibiae. The pronotum is narrowed anteriorly and has an extensive dark central mark usually with lateral lobes and sometimes with metallic reflections. The terminology associated with the surface sculpture is currently misleading enough to risk offering an alternative. The pronotum of *L. bipunctatus* and other *Laccobius* surfaces is often described as shagreened (e.g. Shatrovskiy, 1984) or alutaceous, the old name for *bipunctatus* being *L. alutaceus* Thomson. Shagreenation implies the appearance of polished shark's skin, with raised warts, and alutaceous should refer to the colour and appearance of cracked or wrinkled leather. Neither is suitable for the fine dimpling of the surface, which resembles beaten metal and is clearly visible at a magnification of × 25. The elytra are yellow or grey flecked with black and may be almost entirely black. *Laccobius* paddle strongly below the surface when disturbed and a few species swim freely in open water. In life, the majority stay near to the edge or move within films of water on vertical surfaces.

Females are best identified by association with males. Familiarity with small differences in body shape will ultimately allow both sexes to be identified without dissection of males. However, identification should initially be confirmed by examination of the aedeagophore. Males are easily distinguished by the slightly swollen front tarsi (Fig. 206): they may also have "goggles", a pair of lens-shaped and transparent swellings on the underside of the labrum (Figs 212 and 215). The aedeagophore comprises a median lobe enclosed in a pair of parameres and these structures are in turn enfolded to a greater or lesser extent by an extension of the basal piece.

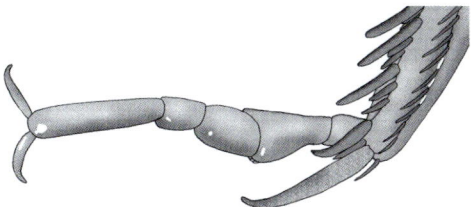

Figure 206. *Laccobius*, male front tarsi.

The species are divided into two small species in the subgenus *Laccobius*, with the elytral punctures organised into rows, and six (Britain) or four (Ireland) larger species in the subgenus *Dimorpholaccobius*, with the elytral punctures rather less organised. The key is constructed bearing in mind that there are almost as many records for *L. bipunctatus* as for the other species together, and that the species most often confused with *L. bipunctatus* is the next commonest, *L. minutus*. It also recognises the overriding value of examining the distinctive male genitalia.

Key 10. The species of *Laccobius*

1. Males: aedeagophore with the parameres much longer than the basal piece (Figs 207, 209 and 210); labrum without obvious goggles. Both sexes: either less than 3.3 mm long or, if larger, with pronotum finely dimpled throughout 2

- Males: aedeagophore with basal piece extending to at least halfway up the aedeagophore (Figs 213, 216, 219, 221 and 223); goggles present (Figs 212 and 215) though sometimes very narrow. Both sexes: 3.2 mm or more long; pronotum either shining between punctures or patchily dimpled 4

2. Male: aedeagophore in dorsal or ventral views distinctive, with parameres curving inwards at their tips (Fig. 207, right); median lobe, viewed from the side, with a small tooth (Fig. 207, left). Both sexes: pronotum dimpled between the punctures; elytral punctures disorganised; length 3.2-3.8 mm ....
.... **2. *Laccobius bipunctatus* (Fabricius)** (p. 70)

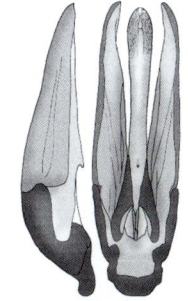

207

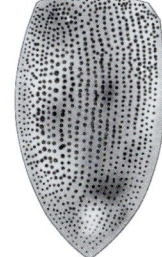

- Male: aedeagophore distinctive with paramere tips straight or splayed (Figs 209 and 210); median lobe without a tooth, viewed from the side. Both sexes: pronotum either dimpled or smooth between punctures; elytral punctures in straggly rows (Fig. 208); length less than 3.3 mm 3

208

3. Males: aedeagophore with narrow, spear-shaped tip to median lobe and parameres straight at tip viewed ventrally or dorsally (Fig. 209). Both sexes: pronotum with fine dimpling between the punctures, sometimes partly effaced but always present; elytral pattern variable and indistinct, the spot at the rear of each elytron being less obvious than in *L. colon* as the punctures within it have some blackening around them (Fig. 208); length 2.4-3.4 mm .... **8. *Laccobius minutus* (Linnaeus)** (p. 72)

- Males: aedeagophore characteristic with expanded tips to parameres (Fig. 210). Both sexes: pronotum smooth between the punctures; elytra each with a distinct white spot on the rear third; spots particularly prominent as the punctures within them lack dark pigment (Fig. 211); length 2.4-3.1 mm .... **7. *Laccobius colon* (Stephens)** (p. 72)

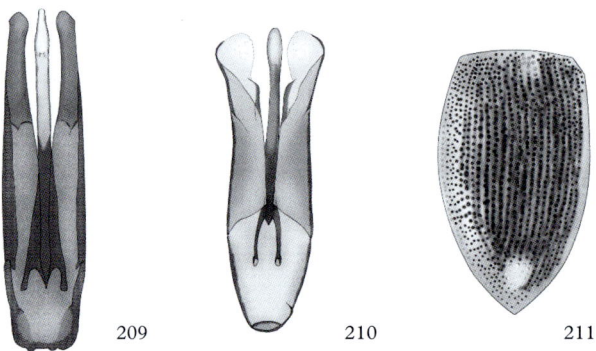

4. Head entirely black and either shining between punctures or weakly dimpled; parameres, as measured along their sides, shorter than the basal piece 5

- Head dimpled and with a pale spot in front of each eye; parameres about the same length as the basal piece 6

    If the upper side is almost entirely metallic purple, leaving only a narrow yellow margin to the pronotum and elytra, proceed to **5. *Laccobius striatulus* (Fabricius)** for its variety *purpurascens* Newbery.

5. Males: goggles narrow (Fig. 212); parameres enfolding median lobe ventrally so that only its tip is visible (Fig. 213) and with their tips only slightly curved inwards. Both sexes: pronotum with only narrow pale margins (Fig. 214); length 3.1-3.8 mm .... **1. *Laccobius atratus* Rottenberg** (p. 69)

- Males: goggles wider (Fig. 215); median lobe exposed more than in *atratus* and long paramere tips distinctively overarching the median lobe (Fig. 216). Both sexes: pronotal dark patch narrower than in *L. atratus*, usually with extensions along its side margins (Fig. 217); length 3.3-4.2 mm .... **6. *Laccobius ytenensis* Sharp** (p. 71)

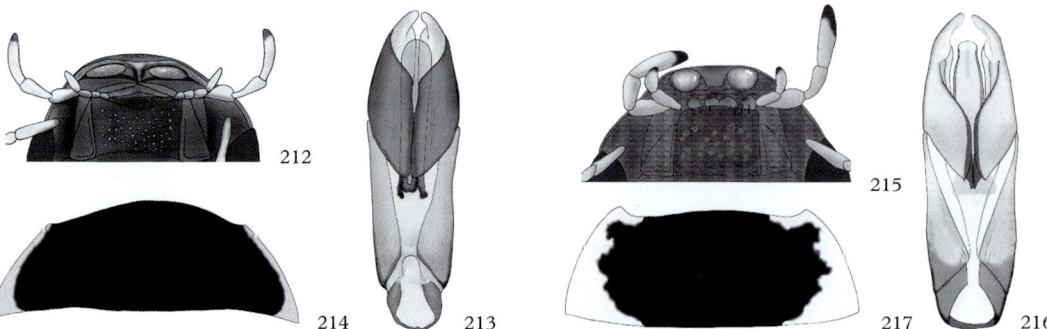

6. Males: mid-femur with a patch of punctures or hairs at the base (Fig. 218). Both sexes: dark patch on pronotum three-quarters or more the width of the pronotum; pronotum either smooth between punctures or only dimpled in part, usually along the front edge ... 7

- Males: mid-femur without closely packed punctures or hairs at the base; aedeagophore with parameres with keeled tips and with their articulation with the basal piece strong and incised at the sides (Fig. 219). Both sexes: dark patch on pronotum just over half the width of the pronotum (Fig. 220); pronotum weakly dimpled throughout .... **3. *Laccobius simulatrix* d'Orchymont** (p. 70)

Known in England from one specimen.

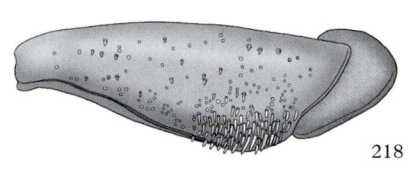
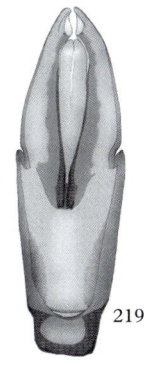
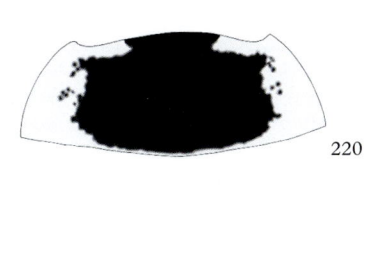

218

219

220

7. Males: aedeagophore broad and typically heavily sclerotised, parameres with a continuous outer profile (Fig. 221). Both sexes: broad oval (Fig. 222), maximum width about 0.6 times the maximum length .... **5. *Laccobius striatulus* (Fabricius)** (p. 71)

- Males: aedeagophore narrow and not so heavily sclerotised; parameres with the outer profile discontinuous (Fig. 223). Both sexes: narrow oval (Fig. 224), maximum width about 0.55 times the maximum length ...... **4. *Laccobius sinuatus* Motschulsky** (p. 71)

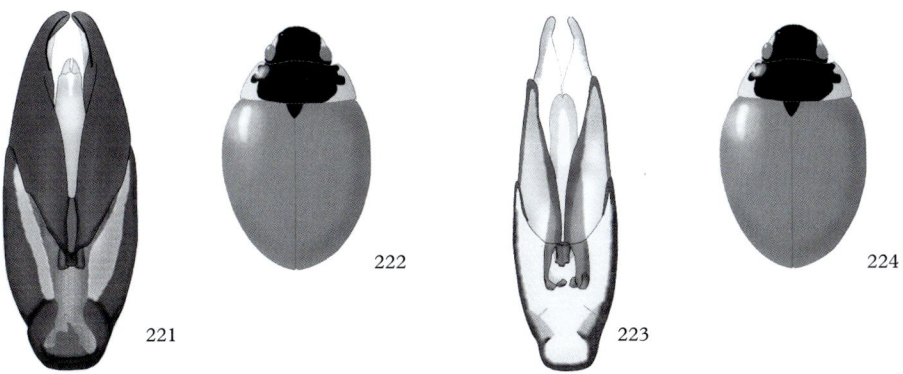

221

222

223

224

Subgenus *Dimorpholaccobius* Zaitzev

1. *Laccobius atratus* Rottenberg Plate 60

Length 3.1-3.8 mm. This species has been initially keyed on the basis of the narrow pale margins to the pronotum: dark specimens of other species have some lobed structure to the dark mark protruding into the margins, an exception being the purple metallic *Laccobius striatulus purpurascens*. The rest of the body of *L. atratus* is dark enough for it to be overlooked as *Anacaena globulus* or *Paracymus scutellaris*, species with which it is often found. Continental specimens may have the body less extensively darkened and *L. ytenensis* may occasionally be almost as dark in Britain and Ireland. Confirmatory characters are provided by the narrow goggles and by the aedeagophore with its long basal piece, resembling that of *L. ytenensis*. The habitat is shallow water, usually seepage, over peat, and typically on moorland or heathland. *L. atratus* occurs around much of the Irish coast. The distribution in Britain is mainly western, north to Kintyre, Arran, Ayrshire and Kirkcudbrightshire, also in the Lake District, most Welsh vice-counties, inland to the Long Mynd in Shropshire, and the south-west. The rest of the southern distribution is on the heaths in Dorset, the New Forest, Berkshire, and Surrey, with 1960s records from Ashdown Forest and adjacent valley bogs in East Sussex, also an early 19th Century record from North Essex. Other records are from northern England, mainly on the North York Moors but extending to Spurn and to the northern Pennines in South Northumberland. *L. atratus* is also known from Jersey. Recorded throughout the year, with peaks in April and August.

2. *Laccobius bipunctatus* (Fabricius) Plate 61

Length 3.2-3.8 mm. Earlier records, when this species was known as *L. alutaceus* Thomson, were confused because of use of the name *bipunctatus* for *L. minutus*. *L. bipunctatus* is the commonest *Laccobius* in Britain and Ireland, usually recognised by the dimpled surface of its pronotum, a character shared with the smaller *L. minutus* and the excessively rare *L. simulatrix*. The aedeagophore of *L. bipunctatus* is like that of *L. colon* and *L. minutus* in having the parameres considerably longer than the basal piece. However, the aedeagophore has a marked tooth visible in side view (Fig. 207) that appears to be unique to this species in the European fauna. *L. bipunctatus* varies considerably in elytral colour and in the extent of darkening of the pronotum, darker specimens being associated with peaty habitats. Small female *L. bipunctatus* can be differentiated from female *L. minutus* by the sharp median keel on the prosternum (Fig. 225), this feature being weaker in *minutus*. The main habitat is muddy shallows wherever these might occur on low ground. *L. bipunctatus* is recorded throughout Ireland and in Britain north to Unst in the Shetlands. Most of the smaller islands have been recorded, an exception being the Scillies. Recorded throughout the year, with peaks in May and August.

Figure 225. *Laccobius bipunctatus*, female prosternum showing sharp median keel.

3. *Laccobius simulatrix* d'Orchymont Plate 62

Length – only known British specimen 3.5 mm, continental authors give 3-4 mm. The British status of this species rests largely on a teneral male taken by Mr E.J. Phillips in a month old clay-bottomed swimming pool at Darvell, East Sussex in July 1982 (Foster & Phillips, 1983). This specimen was originally identified by Dr E. Gentili as belonging to the subspecies *sculptus* d'Orchymont, a taxon subsequently recognised as a distinct species from Turkey and Iraq (Gentili, 1988), and Morocco (Gentili, 2006). Gentili (1977) also noted two specimens in the Brussels Museum labelled "Angleterre" which can be referred to *L. simulatrix* rather than to *L. sculptus*. There is an additional complication that the name *simulator* d'Orchymont was used for a time as an unjustified emendation of *simulatrix*. The aedeagophore of the type of *L. simulatrix* was illustrated by Gentili (2006), and is similar to that of *striatulus*, only narrower with thinner paramere tips and not so heavily sclerotised.

4. *Laccobius sinuatus* **Motschulsky** Plate 63

Length 3.2-4.0 mm. The "narrow oval" shape of *L. sinuatus* is barely different from the "broad oval" of *L. striatulus* but the human eye is good at detecting such differences. The aedeagophore is highly distinctive. *L. sinuatus* lives in the margins of pools, ditches and slow rivers with exposed substratum in lowlands. The water is often base-rich or base-enriched, even polluted, but pristine sites such as landslip pools are also occupied. *L. sinuatus* is frequent in northern East Anglia and other areas draining into the Wash, and north to South Northumberland and extending to the Severn Valley and around London. There are sites scattered around the coast from Kent to South Devon with old records from East Cornwall and North Devon. Welsh records are from Anglesey, Caernarfon, Ceredigion and Glamorgan. *L. sinuatus* is also known from Jersey. There are no Irish or Scottish records and older records may well have resulted from confusion in the use of names for *L. bipunctatus*. Recorded throughout the year, with peaks in April and August.

5. *Laccobius striatulus* **(Fabricius)** Plates 64 & 65

Length 3.5-4.0 mm. This broad and pale species is only likely to be confused with *L. sinuatus*. The robust aedeagophore is heavily sclerotised and dark. *L. striatulus* lives in the margins of drainage ditches, streams and rivers, also occurring on exposed shores of newly created or disturbed pools such as gravel pits and reservoirs. It is well distributed throughout Ireland, Wales, England and southern Scotland but is scarce in the north of Scotland and in some coastal areas further south. Islands with records are Anglesey, Arran, Barra, Brownsea, Islay, the Isle of Man, Mainland Orkney, and the Isle of Wight. Recorded throughout the year, with peaks in May and August. One of the most striking varieties of any water beetle is the metallic purple *purpurascens* Newbery. The most extreme form was originally described from seepage on coastal cliffs on the Devonian Red Sandstone but specimens with varying degrees of purple or coppery colouring have also been reported in riverine habitats north to Fife.

6. *Laccobius ytenensis* **Sharp** Plate 66

Length 3.3-4.2 mm. As forecast by Balfour-Browne (1958) *L. atrocephalus* Reitter has been recognised as a complex within which *L. ytenensis* extends furthest north in Europe. The roof-like attitude of the paramere tips is quite distinctive, and the beetle itself, being relatively narrow with many dark flecks on the elytra, is also recognisable in the field, also because it sometimes flies readily. *L. ytenensis* is mostly associated with hill land at the sides of headwater streams and in valley bogs and other peaty seepage. It is scattered across Ireland with most records on the west coast including the Aran Islands, and Inishbofin and Omey in West Galway. The British distribution is similar, with most records from Sussex to St. Mary's in the Scillies and north to Argyll along the coast, including Anglesey and the Isle of Man. There is an old record from Lewis and other island records from Arran, Cumbrae, Islay and Jura. Records of *L. ytenensis* are much more scattered in the east, from the Wash to the Cheviots, and the few East Anglian records are mostly old. This species is also known from Guernsey and Jersey. Recorded throughout the year, with peaks in May and August.

Subgenus *Laccobius* Erichson

7. *Laccobius colon* (Stephens) Plate 67

Length 2.4-3.1 mm. This species was formerly known as *biguttatus* Gerhardt, both that name and *colon* referring to the distinctive white spots at the rear of the elytra. Other species have spots but not as strongly contrasted to the background colour as in *colon*. The splaying of the parameres is distinctive and, combined with this beetle's small size, there should be no problem in identification. In the field not only are the spots distinctive but the ability to swim deep in open water is striking in a small hydrophilid. The pronotum can be shining or with weak dimpling in patches, but this is never as strong as in *L. bipunctatus* and *L. minutus*. The habitat is base-rich, permanent still water on sparsely vegetated parts of lakes, ponds and fen drains. Common in western Ireland and eastern England, reaching East Perthshire and Angus, with an old record from East Inverness-shire. The only island records are for Anglesey, Bute, Cumbrae, and the Isle of Man. Recorded throughout the year, peaking in May and September.

8. *Laccobius minutus* (Linnaeus) Plate 68

Length 2.6-3.2 mm. *L. minutus* can be confused with *L. bipunctatus*, both having dimpled pronota and similar aedeagophores, but the elytral punctures are in rows in the smaller *L. minutus*. Unaccompanied females can be identified by the elytral rows and by the prosternum having a weak median ridge rather than, as in *bipunctatus*, a keel. Heavily pigmented specimens might be confused with *L. atratus*. The habitat is ponds and drains with some vegetation and exposed substratum. Frequent over much of Ireland and in Britain north to the Orkneys and Outer Hebrides, with records for many other islands, including Lundy, the Scillies and Jersey. Recorded throughout the year, with most records for May and July to September.

Subfamily Sphaeridiinae Latreille

13. *COELOSTOMA* Brullé

The European species are medium-sized, strongly convex, shining black beetles. The eyes are emarginate as in *Sphaeridium* and *Dactylosternum* (Fig. 24) and the antennae are nine-segmented as in *Dactylosternum*. The dorsal surface is evenly punctured with a single stria either side of the elytral suture, reaching about two-thirds towards the front, whereas there are eleven striae on each elytron in *Dactylosternum*.

1. *Coelostoma orbiculare* (Fabricius) Plate 69

Length 4.0-4.8 mm. The generic description notes the possibilities for confusion, the more hemispherical and entirely black body of *C. orbiculare* also distinguishing it from *D. abdominale*. Another possibility is the other European species, *C. hispanicum* (Küster), which reaches north to the Loire valley; this is similar in size but with paler palps and tarsi. *C. orbiculare* is typical of moss in floating rafts of vegetation but also occurs in edges of ponds and ditches, usually in association with mosses, sometimes with rotting vegetation. Frequent across most of mainland Ireland and Britain, and often recorded on smaller islands:- Anglesey, Great Blasket, Brownsea, Clare, most Clyde Isles (Arran, Bute, Holy Island and Little Cumbrae), Guernsey, Jersey, Lismore, the Isles of Man and May,

Rathlin, Skokholm, Tresco, and the Isle of Wight. In the north of Scotland *C. orbiculare* reaches the north Caithness coast and is known from most Hebridean islands north to North Uist. Recorded through the year, peaking over May and June.

14. *DACTYLOSTERNUM* Wollaston

The only European species is like a longer and flatter version of *Coelostoma* with elytral striae, but in the field, it resembles more an *Aphodius* or even an unusually elongated histerid.

1. *Dactylosternum abdominale* (Fabricius) Plate 70

Length 3.8-5.0 mm. In addition to the characters mentioned above *D. abdominale* is typically black with narrow red margins and a reddish underside. The habitat is almost any kind of decaying organic matter, the first find in England being at the edge of a silage clamp and the second in a plastic bin used to compost vegetable waste (Welch, 2006). The English habitats include a tussock on a chalk down, horse dung, a rotting log and a rotted bracket fungus. First reported in Dorset in 2003 (Allen, 2004), this species is additionally known from South Somerset, East Kent, Bedfordshire and West Norfolk. Recorded so far from August to November. This species has become almost cosmopolitan, originally being described from tropical America.

15. *CERCYON* Leach

This is the main genus of small species within the British and Irish Sphaeridiinae. The sides of the head are strongly contracted in front of the eyes, which are regularly rounded as opposed to the indented eye margins of *Coelostoma*, *Dactylosternum* and *Sphaeridium*. The characters of the raised parts of the mesosternum and the metasternum (the latter descriptively known as a "tablet", with a "metasternal process" on its front) set *Cercyon* apart from *Cryptopleurum* and *Megasternum*. The strongly convex appearance and the prominent antennae the same length as the maxillary palps should distinguish aquatic members of the genus from other water beetles:

Aquatic *Cercyon* cannot swim, walking just beneath the surface when disturbed from marginal vegetation and litter. Insight into the ecology of *Cercyon* is provided by the occurrence of European species in North America. A few species are genuinely Holarctic, as evidenced by the occurrence of subfossil material in America, but most European species have become more recently established in North America (Šmetana, 1978), often on both seaboards. However, none of the microreticulate species, which are aquatic, has expanded in this way, emphasising the importance of human agency in distributing the species that can survive in manure. Mass flights of insects recorded by Southwood & Johnson (1957) included eight species of *Cercyon*, *C. marinus* being the only aquatic representative. The Oriental *C. laminatus* is undoubtedly introduced to Britain and *C. nigriceps* may also have originated in India or the Far East. Associations of *Cercyon* species with birds' nests in eastern Europe (Ryndevich & Lundyshev, 2005; Ryndevich, 2008; Buczyński & Tończyk, 2010) are recorded in detail here, given that this kind of investigation is almost impossible in western Europe. The bird names follow Dudley *et al.* (2006). Of the twenty-two species of *Cercyon* known to occur in Britain seven are more likely to be found in or at the edge of freshwater rather than on land, the equivalent numbers for Ireland being eighteen and five. The key is constructed to allow recognition of the wetland species separately from the rest but recognising the possibility that terrestrial species can occasionally occur in or near water, often in flood refuse.

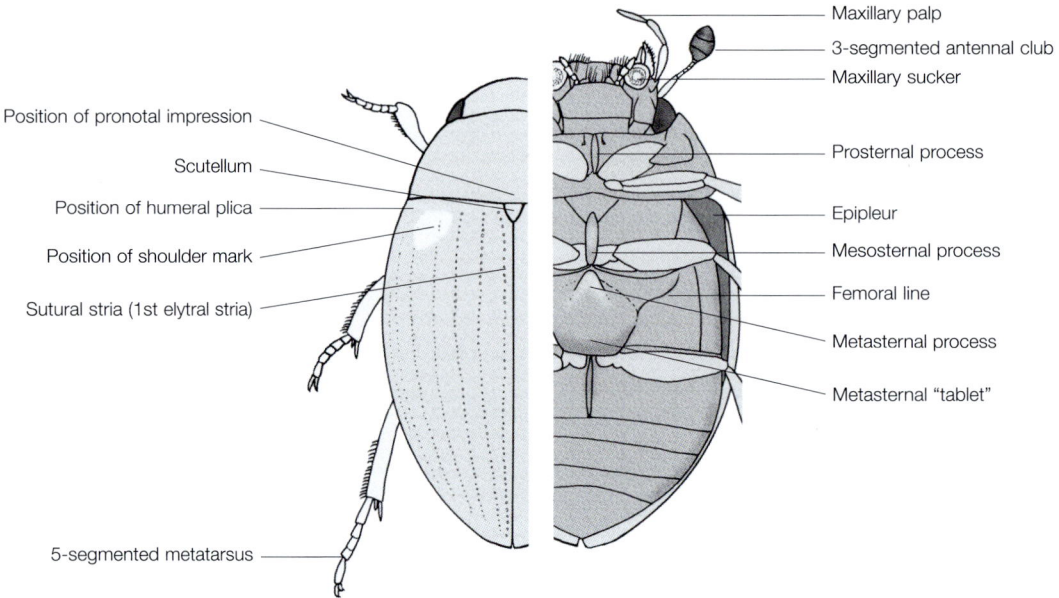

Maxillary palp
3-segmented antennal club
Maxillary sucker

Position of pronotal impression

Prosternal process

Scutellum

Position of humeral plica

Epipleur

Position of shoulder mark

Mesosternal process

Sutural stria (1st elytral stria)

Femoral line

Metasternal process

Metasternal "tablet"

5-segmented metatarsus

Figure 226. Features of a *Cercyon*.

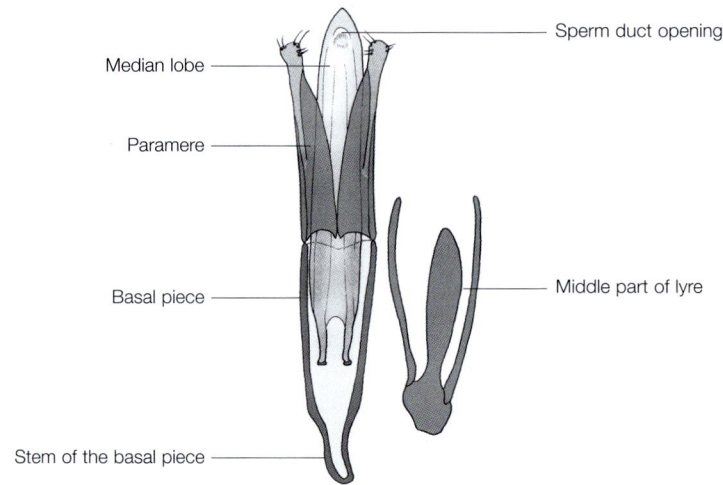

Sperm duct opening

Median lobe

Paramere

Basal piece

Middle part of lyre

Stem of the basal piece

Figure 227. Features of the male genitalia of a *Cercyon*.

Most *Cercyon* can be identified by their colour patterns but there will always be a few individuals that do not conform, either being extremes of natural variation, or pale when recently emerged – or even having been a long time in a museum cabinet. Otherwise, identification largely hinges on thoracic features, such as the width of the mesosternal process. On the elytra it should be noted that there are 11 punctured striae, counting in the sutural stria, but that there are finer punctures on the interstices and these too may be organised into rows. The male genitalia provide a limited range of distinguishing features. The sex of a *Cercyon* can be determined externally by detection of sucker discs on the male maxillae but the tips of the genitalia, typically of the pair of cerci in the female, may be visible. Dissection can be aided by adhering the specimen to some wall-mounting putty (Blu-Tack®, etc.). Two reviews of European species (Šmetana, 1978; Huijbregts, 1982) depict the male genitalia further dissected into its constituent parts but, despite detail being obscured, it seems best to examine the aedeagophore intact, as the extent to which the median lobe protrudes beyond the parameres, or is enclosed by them, is also characteristic. It is important to mount the parameres in a medium such as DMHF immediately after dissection as their tips are delicate and may become distorted if dried out. The preparation could usefully be on a transparent strip of acetate rather than on the mounting card itself. Detail such as the tips of the parameres may not be visible except by transmitted light. The other part of the male genitalia is a three-pronged structure adhering to the underside of the aedeagophore, referred to sometimes as a lyre (Fig. 227): it is often depicted in reviews of the genus but has limited use as an aid to identification.

Four subgenera are currently recognised for the British species, three of them occurring in Ireland. They are of limited value in identification, the majority of species being assigned to the nominate subgenus.

Key 11. The species of *Cercyon*

1. Beetles found in or at the edge of water, but not beach driftlines 2

- Beetles found in other habitats – dung, vegetable debris, rotting seaweed, nests, etc. 8

2. In side view, the elytra and pronotum are separately domed (Fig. 228), the pronotum being more domed than the front of the elytra; rear of pronotum with a groove just above and slightly shorter than the scutellum (Fig. 229); mesosternal process very narrow (Fig. 230); aedeagophore Fig. 231; length 2.6-3.2 mm .... Subgenus *Dicyrtocercyon* Ganglbauer
   20. *Cercyon ustulatus* (Preyssler) (p. 89)

- Elytra and pronotum forming a smooth curve; rear edge of pronotum without a groove ... 3

   C. impressus, which is occasionally found in water, has a groove on the pronotum as in *C. ustulatus*.

228

229

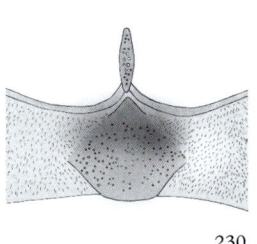

230

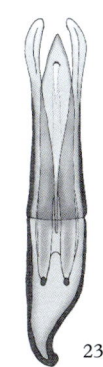

231

3. In contrast to the smooth pronotum, the elytra have a net-like reticulation, visible at × 30 on a dry beetle (Fig 232) and in which individual reticulations are no larger than a puncture; length 1.6-2.3 mm 4

   These small beetles are black, often with the rear of the elytra pale. Variations in the colouring of the appendages are too great to make them dependable for identification.

- Elytra without net-like reticulation, or rarely with a loose reticulation of fine lines encompassing several punctures (see *C. analis*, Fig. 297); length 1.7-4.2 mm 7

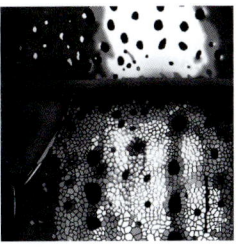

232

4. The mesosternal process separated from the metasternal "tablet by a distinct gap (Figs 233 and 236) 5

- Mesosternal process touching the tablet or at least overlapping it (Figs 239 and 241) 6

If in doubt about this character view the structures from the side or look for the tooth on the front of the metasternal tablet in *C. tristis* (Fig. 236)

5. Front of metasternal tablet rounded (Fig. 233); all rows of punctures remaining strong to the tip of the elytra (rear/side view at × 25) (Fig. 234); elytra microreticulate but shining almost as much as the pronotum; aedeagophore with pointed parameres and a broad extension to the basal piece (Fig. 235); length 1.7-2.4 mm .... **5. *Cercyon granarius* Erichson** (p. 84)

- Front of metasternal tablet drawn out into a tooth (Fig. 236); primary punctures in the middle rows of each elytron becoming progressively disorganised and feeble towards the rear, largely disappearing in the area circled (Fig. 237); elytra dull relative to pronotum; aedeagophore with truncate tips to the parameres (Fig. 238); length 1.7-2.3 mm .... **18. *Cercyon tristis* (Illiger)** (p. 88)

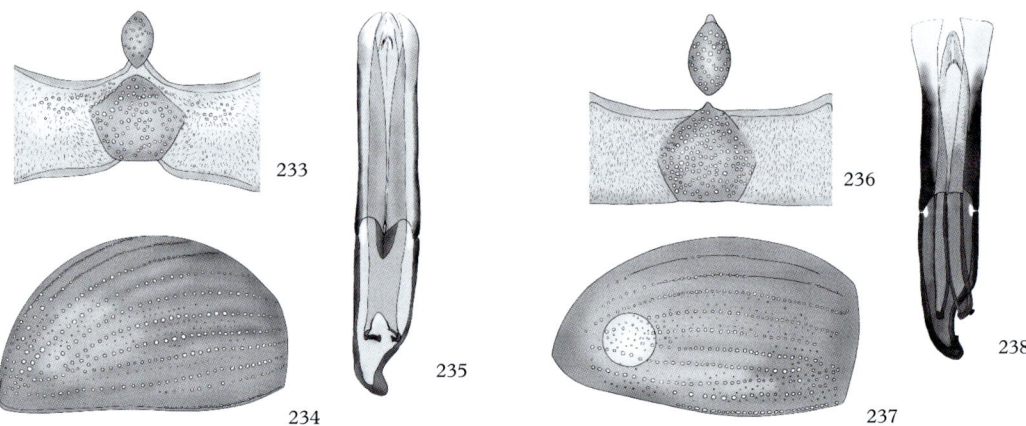

233   236   235   234   237   238

6. Mesosternal process more than three times as long as wide (Fig. 239); last segment of palps darkened; aedeagophore with median lobe well short of the slightly expanded, pointed and clasping parameres (Fig. 240); length 1.6-2.2 mm .... **3. *Cercyon convexiusculus* Stephens** (p. 83)

- Mesosternal process broader, about 2½ times longer than wide (Fig. 241); palps uniformly reddish brown; aedeagophore with the median lobe as long as the parameres, sometimes protruding beyond them (Fig. 242); length 1.6-2.0 mm .... **16. *Cercyon sternalis* (Sharp)** (p. 88)

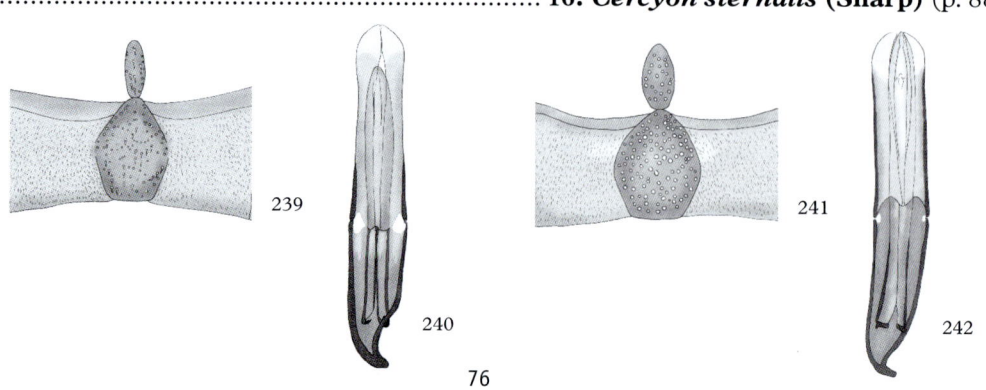

239   241   240   242

7. Pale side borders of the elytra extending near to the front edge (Fig. 243), sometimes even reaching it; mesosternal process nearly 3 times as long as broad and surface flat (Fig. 244), not concave; aedeagophore slender, its parameres with wide and rounded transparent tips covering the tip of the median lobe (Fig. 245); length 2.2-3.4 mm .... **10. *Cercyon marinus* Thomson** (p. 86)

- Pale side borders extending only half way up the elytra (Fig. 246); mesosternal process less than twice as long as wide (Fig. 247) and with its surface slightly concave; aedeagophore unusually large and robust with parameres slightly splayed, their tips being blunt and transparent, and leaving the tip of the median lobe exposed (Fig. 248); length 2.2-3.0 mm .... **2. *Cercyon bifenestratus* Küster** (p. 83)

If neither of the descriptions in couplet 7 appears appropriate, continue to couplet 8.

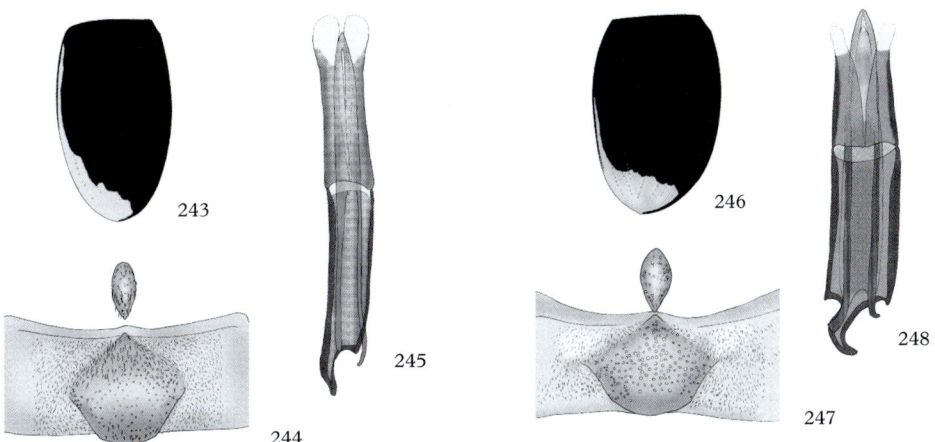

8. Spearhead-shaped central part of the mesosternum forming a sharp keel on its ventral edge (Fig. 249); distinctive rather flat and large species with a black head contrasting with the paler pronotum and elytra (Plate 93); scutellum narrow (Fig. 250); aedeagophore with the median lobe longer than the parameres, which have outwardly deflected narrow tips (Fig. 251); length 3.2–4.0 mm .... Subgenus *Paracycreon* d'Orchymont
22. *Cercyon laminatus* Sharp (p. 90)

- Mesosternum with a raised process which may be wide or narrow, but not sharply keeled; if a large species the head and pronotum are similarly dark; elytral suture often darker than the rest of the elytra; scutellum broader than in Fig. 250 9

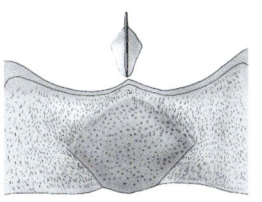

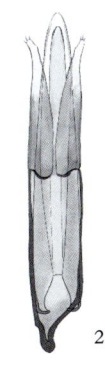

9. Rear of the raised mesosternal process fitting into a V-shaped slot into a forward extension of the metasternal process (Fig. 252); elytra strongly narrowed towards the rear (Fig. 253); aedeagophore Fig. 254; length 1.7-2.8 mm .... Subgenus *Paracercyon* Seidlitz
21. *Cercyon analis* (Paykull) (p. 89)

- Raised mesosternal process meeting the metasternum at a point or separated from it by a small gap (e.g. Fig. 287), or overlapping but without engaging in a slot (e.g. Fig. 289); elytra not so strongly contracted at the rear 10

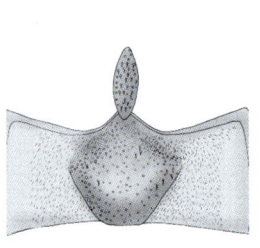
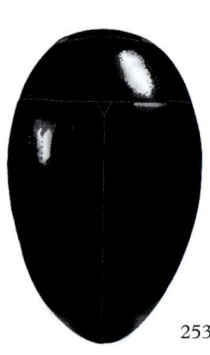

252

253

254

10. Metasternum with distinct "femoral lines" running either side of the metasternal tablet (e.g. Figs 259 and 262) 11

- Metasternum without femoral lines (e.g. Figs 274 and 277) 16

11. Femoral lines not reaching the front margin of the metasternum (Figs 256, 259, 262 and 265); length 2.3-4.0 mm 12

- Femoral lines reaching the front margin (Figs 268 and 271); length 1.4-2.0 mm 15

12. Pronotum with a dent or short groove just in front of the scutellum (Figs 255 and 258), sometimes no more than a weak depression with a gap amongst the punctures 13

The "*impressus*" groove or depression can be difficult to see but it is important in setting these species apart. The specimen should be thoroughly cleaned and viewed in diffuse light, rotating it to catch the light on the area of the impression.

- Pronotum with the punctures of the rear margin not interrupted by any median depression or groove 14

13. Length 3.0-4.0 mm; pronotum with a short dent just in front of the scutellum, sometimes accompanied by a groove, often weak and marked out by a gap in the punctures (Fig. 255); median lobe of aedeagophore broad but with a narrowly pointed tip and projecting beyond the round tips of the parameres (Fig. 257); mesosternal process and metasternum Fig. 256 .... **7. *Cercyon impressus* (Sturm)** (p. 85)

-   Length 2.5-2.9 mm; pronotum with a very narrow groove longer than the distance between the outer edges of two adjacent punctures in its vicinity (Fig. 258); median lobe very narrow and not projecting beyond the deflected paramere tips (Fig. 260); mesosternal process and metasternum Fig. 259 .... **1. *Cercyon alpinus* Vogt** (p. 83)

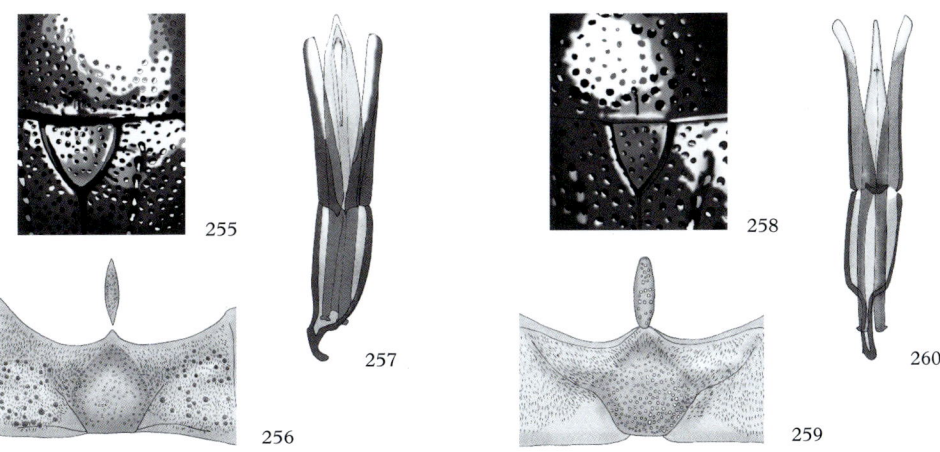

14. Elytra red with a sharply defined triangular black band and black epipleurs (Fig. 261); aedeagophore Fig. 263; length 2.3-3.0 mm; mesosternal process and metasternum Fig. 262 .... **11. *Cercyon melanocephalus* (Linnaeus)** (p. 86)

-   Elytra with the base and part of the sutural stria darkened to form a T-shaped, never triangular, mark (Fig. 264), or, if more extensively darkened, then the T pattern is visible with light transmitted through the elytron; epipleurs brown; aedeagophore Fig. 266; length 2.5-3.2 mm; mesosternal process and metasternum Fig. 265 ....
.... **6. *Cercyon haemorrhoidalis* (Fabricius)** (p. 84)

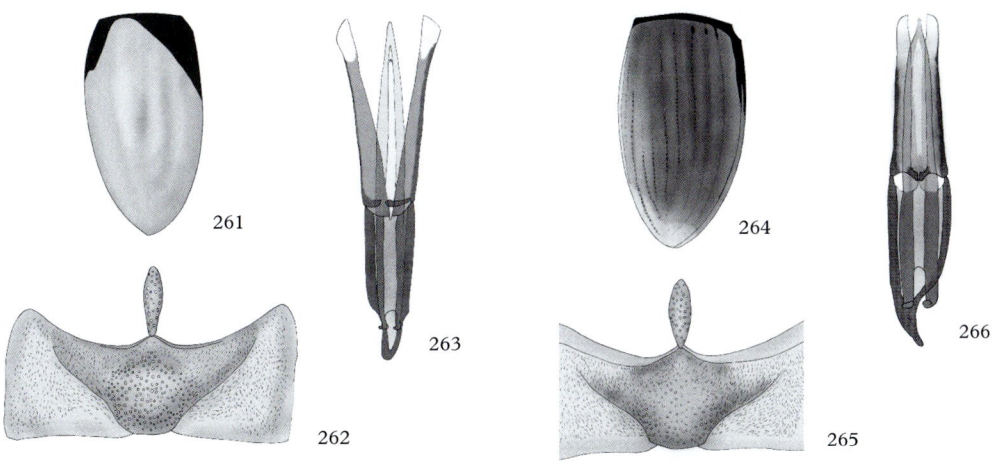

15. Lateral margin of pronotum rounded in rear third in side view (Fig. 267); raised rim of pronotal margin continued along the rear edge; innermost elytral striae reaching the front; median lobe of aedeagophore narrow (Fig. 269); length 1.4-2.0 mm; mesosternal process and metasternum Fig. 268 .... **12. *Cercyon nigriceps* (Marsham)** (p. 86)

- Lateral margin of pronotum bluntly angled just beyond the midpoint and straight in the rear third (Fig. 270); raised rim of the pronotum confined to the sides; innermost striae weak or absent at the front of the elytra; median lobe of aedeagophore broad (Fig. 272); length 1.2-1.8 mm; mesosternal process and metasternum Fig. 271 .... **14. *Cercyon pygmaeus* (Illiger)** (p. 87)

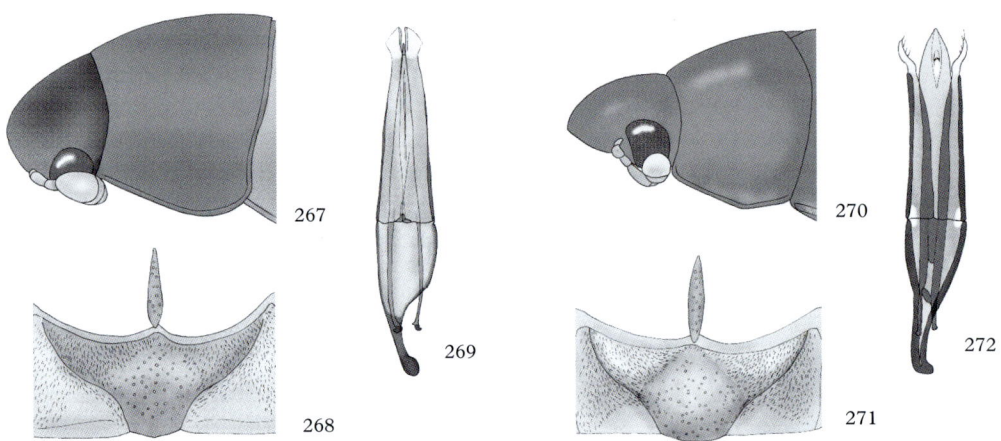

16. Front tibia obliquely cut across at the tip, with a prominent, blunt spine behind which there is a gap (Fig. 273); aedeagophore Fig. 275; length 2.6-3.0 mm; mesosternal process and metasternum Fig. 274 .... **9. *Cercyon littoralis* (Gyllenhal)** (p. 85)

The commonest *Cercyon* beneath rotting seaweed on the beach.

- Front tibia rounded at its tip 17

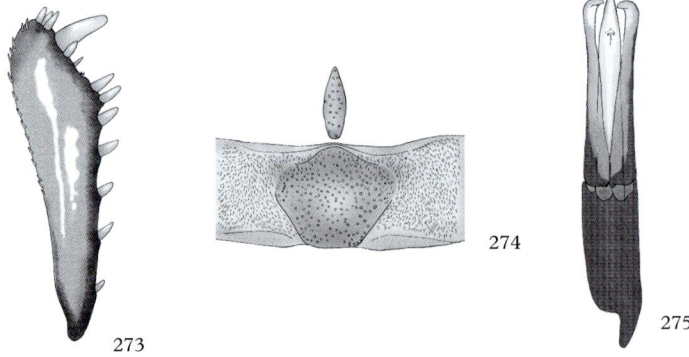

17. Viewed from the side, the pronotal edge is sinuate just before the hind margin (↖ Fig. 276); viewed from above the hind edge of the pronotum is narrower than the front of the elytra (Plate 74); aedeagophore Fig. 278; length 1.8-2.6 mm; mesosternal process and metasternum Fig. 277 .... **4. *Cercyon depressus* Stephens** (p. 84)

Another seaweed species.

- Sides of the pronotum curving up gradually to the hind edge (Fig. 279), which is as wide as the front of the elytra 18

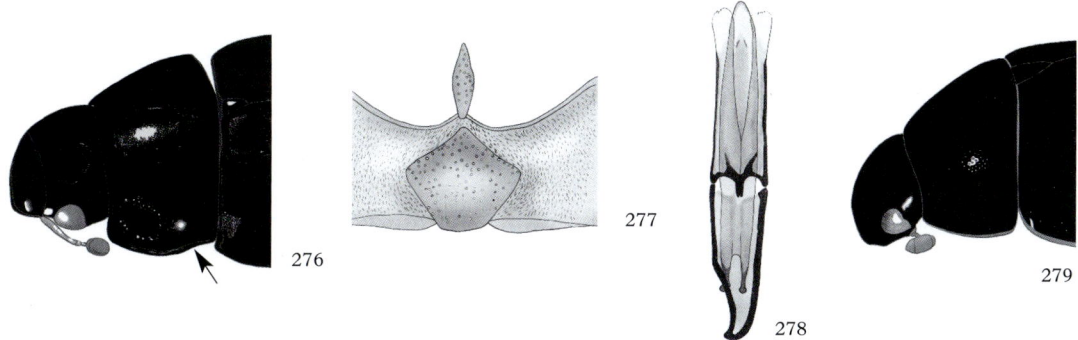

276   277   278   279

18. Elytral margin with a small oblique depression just behind the anterior angle (Fig. 280 ➤); maxillary palps black; aedeagophore with basal piece longer than parameres (Fig. 282); length 3.3-4.2 mm; mesosternal process and metasternum Fig. 281 ....
.... **13. *Cercyon obsoletus* (Gyllenhal)** (p. 87)

- Elytral margin smooth; maxillary palps paler; aedeagophore with the parameres longer than the basal piece; less than 3.4 mm long 19

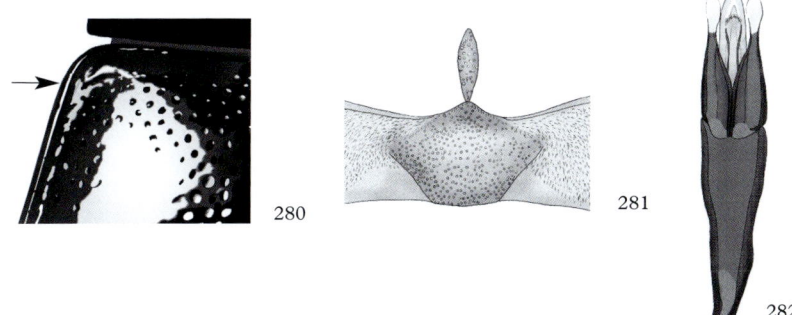

280   281   282

19. Mesosternal process narrow (Figs 284, 287, 289 and 291); elytra variously coloured, but without a sharply defined apical spot extending along the elytral edge 20

- Mesosternal process wider (Figs 244 and 247); elytra with a sharply defined apical spot extending along the elytral edge .... **return to 7**

20. Elytra usually yellow with a sharply defined dark spot in the middle (Fig. 283); punctures of the 9th and 10th elytral striae equally distinct, though those of the 10th fewer and stopping just after halfway along the elytra; aedeagophore Fig. 285; length 2.4-3.4 mm; mesosternal process and metasternum Fig. 284 ...... **19. *Cercyon unipunctatus* (Linnaeus)** (p. 89)

- Elytra without a spot as above; punctures of the 10th elytral stria finer than those of the 9th (Fig. 286) 21

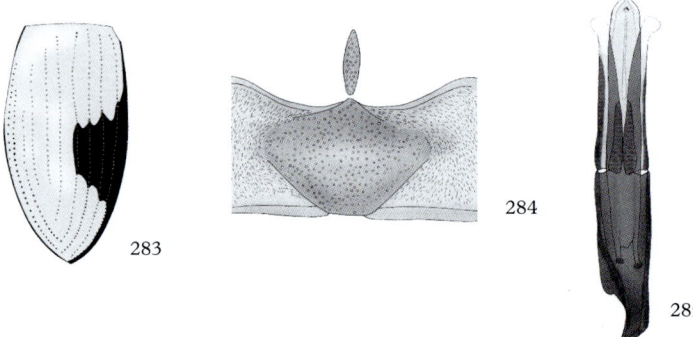

21. Elytra with a weakly defined yet distinctive pattern, with the suture paler over the front half and dark over the rear, surrounded by large dark areas leaving the elytral shoulders and the rear paler (Fig. 286); aedeagophore Fig. 288; length 2.5-3.2 mm; mesosternal process and metasternum Fig. 287 .... **8. *Cercyon lateralis* (Marsham)** (p. 85)

- Elytral pattern not as above; smaller species, not exceeding 2.6 mm long 22

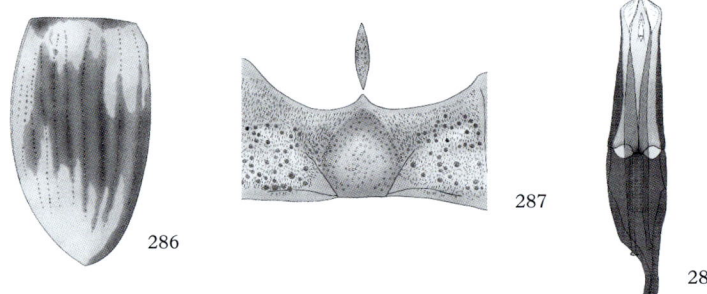

22. Pronotum entirely black; elytra either dark, reddish with a weak version of the pattern of *C. melanocephalus* (Fig. 261), or entirely pale; antennal club pale; aedeagophore Fig. 290; length 1.6-2.3 mm; mesosternal process and metasternum Fig. 289 ....
.... **17. *Cercyon terminatus* (Marsham)** (p. 88)

- Pronotum black with lateral margins paler; elytra yellow; antennal club brown; aedeagophore Fig. 292; length 2.0-2.6 mm; mesosternal process and metasternum Fig. 291 .... **15. *Cercyon quisquilius* (Linnaeus)** (p. 87)

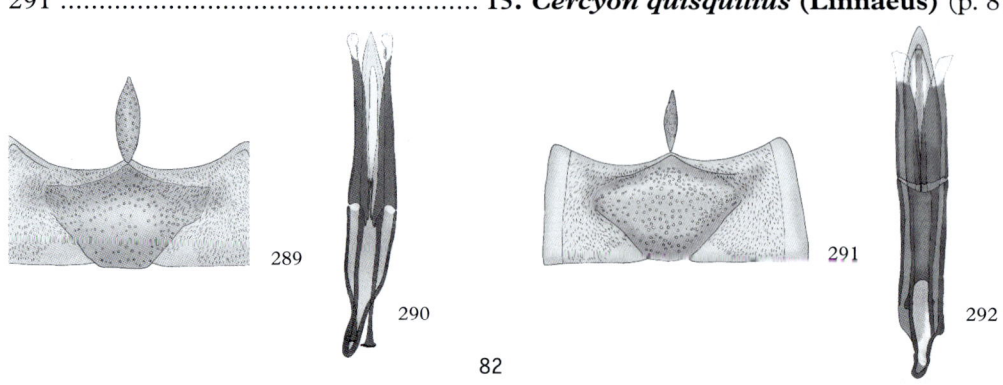

Genus *CERCYON* Leach
Subgenus *Cercyon* Leach

1. *Cercyon alpinus* Vogt Plate 71

293

Length 2.5-2.9 mm. In the key *C. alpinus* is in the same couplet as the larger *C. impressus* on the basis of the presence of the median pronotal groove and the presence of femoral lines not reaching the front edge of the metasternum. *C. alpinus* has the same elytral colouring as *C. impressus* and also as the darker form of the similarly sized *C. haemorrhoidalis*. The elytra are largely black with an extensive dark red mark at the rear, sometimes extending to a pale spot near the shoulder and on the outer edge but leaving much of the suture dark (Fig. 293). There are pale variants on this, just as there are in the other two species, with a large black triangle around the suture, but the epipleurs of *C. alpinus* and *C. impressus* are dark whereas they are orange in *haemorrhoidalis*. *C. alpinus* was originally described from cow dung but it has been found in Scotland in the dung of red deer and in England in horse dung. The British sites are partly wooded. The beetle has also been found in pitfall traps. The Scottish sites are the Forest of Mar and Braemar in South Aberdeenshire and, in East Inverness-shire the Abernethy Estate (also in Moray) (Owen 1994) and Guisachan Forest. The two English sites are in West Gloucestershire (Bratton, 1998). Recorded from May to September.

2. *Cercyon bifenestratus* Küster Plate 72

Length 2.2-3.0 mm. The extent of elytral colouring is sufficiently constant to distinguish this species from *C. marinus*. The aedeagophore is also distinct, and the mesosternal process has a weakly concave surface. The habitat is a flooded and exposed substratum, usually sand or mud, with a small amount of organic debris, and can be provided in disturbed sites such as quarries and flood relief structures. There are also reports of association with the faeces of geese and swans, also under bark covered with rabbit pellets (Whitehead, 2005), and in the nests of black-headed gulls and mute swan. *C. bifenestratus* is so far only recorded from England, in North Somerset, East Sussex, East and West Kent, Surrey, Hertfordshire, Middlesex, Berkshire, East and West Suffolk, West Norfolk, Cambridgeshire, East Gloucestershire, Worcestershire, Warwickshire, North Lincolnshire, Leicestershire, and Nottinghamshire. Recorded in January, March and in May to September, peaking in August.

3. *Cercyon convexiusculus* Stephens Plate 73

Length 1.6-2.2 mm. There is some uncertainty about the identity of type material of this species, and it might be noted that the name *Cercyon intermixtus* Sharp, 1918 is available. *C. convexiusculus* is the only microreticulate species with a narrow mesosternal process. In addition to characters used in the key it should be noted that *C. sternalis* has secondary punctures on the elytral intervals so fine as to be barely visible at x 80 whereas the punctures of *C. convexiusculus* are distinct, each about the size of one mesh of the microreticulation. Its habitat is wet litter in fens, often in association with tussocks. Ryndevich & Lundyshev (2005) recorded it from nests of a black-headed gull, a tufted duck and a song thrush. This is the commonest of the reticulate species in England, avoiding high ground, absent from the south-west and north-west, and requiring confirmation north of the Tyne valley on the east coast. It is extremely rare in Scotland, with records from the Carrick Ponds and Carlingwark Loch in Kirkcudbrightshire and the Lake of Menteith in West Perthshire. *C. convexiusculus* is infrequent in Wales away from coastal marshland, with records from Anglesey, Denbighshire, Flintshire, Glamorgan, Monmouthshire, Montgomeryshire, and Radnorshire. The Irish distribution is lowland, ranging from East Donegal in the north to Sherkin Island, West Cork, in the extreme south. There is also a record for Jersey requiring confirmation: this is based on a manuscript entry by Mr K.C. Side, but his collection does not contain a specimen. Recorded throughout the year, peaking in May.

4. *Cercyon depressus* **Stephens** Plate 74

Length 1.8-2.6 mm. This is a diminutive version of the commoner *C. littoralis*, but is more parallel-sided. The fore tibiae lack the gap between the spines easily seen in *C. littoralis*. This species is usually black with the lateral margins and the rear of the elytra yellowish, sometimes also with a small pale shoulder spot. Brown specimens always have the suture black. *C. depressus* is found under heaps of decaying wrack on the beach. In Ireland it was in the past (Haliday, 1855) noted that it can occur on more shingly beaches than *C. littoralis* and Šmetana (1978) noted its occurrence under stones. Scattered around the coast of Britain, with few records from the east coast of England. Its northernmost records are from the Brough of Birsay and Glims Holm in the Orkneys, other island records including Alderney, Anglesey, Lundy, the Isle of Man, Raasay, Rum, Tresco, South Uist, and Whiddy Island in West Cork. *C. depressus* is also known from Lough Neagh in Londonderry. Recorded from March to October with a peak in June.

5. *Cercyon granarius* **Erichson** Plate 75

Length 1.7-2.4 mm. This is the rarest of the microreticulate species of *Cercyon*. The reticulation is so weak that the elytra are almost as shining as the pronotum. The broad mesosternal process is shared with *C. sternalis* and *C. tristis*, but it can be distinguished from *C. sternalis* by the presence of a gap between that process and the metasternum (Fig. 233), and by the front of the metasternum being blunt, not drawn out into a process to meet the mesosternum as in Fig. 236 in *C. tristis*. These characters provide an important check on the character provided by the intensity of the rows of punctures on the elytra as depicted in the key (Fig. 234). The main habitat, which appears to be amongst floating vegetation, is provided by rich lowland fens, also occasionally by rivers and ponds. Winter records for this species are from flood debris. Ryndevich & Lundyshev (2005) recorded it from the nest of a black-headed gull. Recorded in England from North Somerset, South Wiltshire, East and West Kent, East and West Sussex, Surrey, East Suffolk, East and West Norfolk, Oxfordshire, and Warwickshire. *C. granarius* is also known from Jersey. Recorded from April to August and in November, January and February.

6. *Cercyon haemorrhoidalis* **(Fabricius)** Plate 76

Length 2.5-3.2 mm. This species is easily recognised when the elytral colouring is reduced to the dark T-shaped mark covering the base. Even when the elytra are dark the intensity of the T can still be detected. The habitat is rotting organic matter from dung to compost, carrion and sap. Ryndevich & Lundyshev (2005) recorded it from the nest of a black-headed gull. It is common in lowland Britain, less so on the coast and with other gaps that may simply reflect a lack of recording effort. It ranges from the Scillies, in Bryher and St. Mary's, to St. Kilda, the Shetlands, and the Faroes, but appears to be absent from around the Wash and the Broads. The Irish distribution is mainly coastal except in the north, where it appears to be more common. Other island records include Guernsey, Jersey, Sark and the Isle of Man. Recorded throughout the year, peaking in May.

7. *Cercyon impressus* (Sturm) Plate 77

Length 3.0-4.0 mm. This beetle shares with *C. ustulatus* the groove at the rear of the pronotum but differs from it in having femoral lines, in lacking the double-domed profile, and in having the median lobe protruding beyond the parameres. Its colour is generally dark with a vaguely defined posterior dark red mark that may be so extensive as to leave a large black triangle, although never quite as clearly defined as in *C. melanocephalus*. The habitat is decaying organic matter under almost any condition from exposed riverine sediment to dung, usually from cattle or sheep, but also from red deer. *C. impressus* is widely distributed with clustering on maps indicating the activity of specialist recorders. A prominent feature of its distribution in England is that it is barely recorded from the east and south coasts, being almost absent from East Anglia, and yet is also scarce on high ground. It is frequent in Wales and southern Scotland and scattered across the Highlands north to Ross and Moray. Island records are for Anglesey, Arran, Bardsey, Jersey, Lismore, the Isle of Man, Raasay, Rum, the Scilly Islands Bryher and St. Mary, Seil and Tiree. The scattered Irish distribution indicates that it is frequent throughout, including lowland and mountainous areas. Recorded throughout the year, with peaks in May and October.

8. *Cercyon lateralis* (Marsham) Plate 78

Length 2.5-3.2 mm. This large species should be recognised by its elytral pattern. A feature apparently unique to this species is the pitting of the outer parts of the metasternum (Fig. 287): these pits are difficult to see, often being only visible as partings in the covering of hair. The habitat is any decaying matter including dung, compost and carrion: Skidmore (1991) specifies horse dung as opposed to that of cattle and sheep. Ryndevich (2004) notes occurrence in the dung of horses, cattle, European Bison (*Bison bonasus* (L.)), red deer (*Cervus elephus* L.) and other mammals, and rotting plant matter including the stinkhorn fungus (*Phallus impudicus* L.), the bracket fungi *Piptoporus betulinus* (Bulliard) and *Polyporus squamosus* (Hudson). Ryndevich & Lundyshev (2005) recorded it from the nests of a song thrush and a European honey-buzzard. Frequent across lowland Britain, and in the north and south-west of Ireland. The northernmost records are from Mainland Orkney, Coll, Canna, Raasay and Tiree, and other islands with records are Anglesey, Bute, the Isles of Man and May, and Jersey. Recorded in all months except December, peaking in May and September.

9. *Cercyon littoralis* (Gyllenhal) Plates 79 & 80

Length 2.5-3.3 mm. The body shape of this and *C. depressus* is more elongate than other *Cercyon*, and the shape of the fore tibiae should put identification beyond doubt. The aedeagophore is robust, but with the tips of the parameres flimsy and turned inwards to varying degrees. This is the larger of the two species found in wrack on the beach, and it is usually the commoner of the two. Pale forms of *C. littoralis* have attracted names, *C. littoralis* var. *binotatum* Stephens being the most distinctive, with the upper surface yellow except for the head and a pair of black spots variously developed towards the rear of the elytra. *C. littoralis* var. *ruficollis* Schilsky is a more generally reddish form. *C. littoralis* is to be found almost anywhere with the habitat of decaying wrack on a solid beach, i.e. sand or rock, but not shingle. This species extends from the Shetlands and St. Kilda south to the Scillies, and is known from Alderney and Herm. Pale forms are often found in populations all over Britain and Ireland, and also north to the Faroes (West, 1930). Recorded from January to October, with peaks in May and July.

10. *Cercyon marinus* **Thomson** Plate 81

Length 2.2-3.4 mm. *C. marinus* is a brightly marked and shiny species, separable from the rarer *C. bifenestratus* by the characters provided in the key. The habitat is vegetated muddy or mossy edges of permanent and still waters, rarely on riverbanks. Ryndevich & Lundyshev (2005) recorded it from the nests of mute swans and black-headed gulls, and *C. marinus* has also been found in association with mallard faeces. It is common in the English Midlands, mostly coastal elsewhere in England, found in the Somerset Levels but not reported further west except for St. Mary's and Tresco in the Scillies. There are records in Wales from Anglesey, Ceredigion, Meirionydd, Monmouthshire, and Skokholm in Pembrokeshire. *C. marinus* is known in Scotland north to Fife, appears to be absent from the Highlands and the Hebrides but then is known from Orkneys and Shetlands, consistent with its reported occurrence in the Faroe Islands. The only other Scottish island from which it is recorded is Little Cumbrae, a frequent association with coastal extremities being emphasised by records from the Isle of Whithorn, Wigtownshire, and Ardmore Point in Dunbartonshire. The Irish distribution is largely coastal from Whiddy Isle in West Cork in the extreme south to Garry Bog in the north of Antrim, but the most noticeable feature is its frequency around Lough Neagh. *C. marinus* is also known from Jersey. Recorded throughout the year, with peaks in May and July.

11. *Cercyon melanocephalus* **(Linnaeus)** Plate 82

Length 2.3-3.0 mm. This is one of the most easily recognised *Cercyon* in the field, its red elytra having a distinct black triangle around the scutellum and dark epipleurs. The pattern is constant. Other species may have similar but more obscure patterns. The aedeagophore has no distinctive features – the parameres are longer than the basal piece and may be found splayed or clasping the median lobe, which may protrude beyond them. Found in all kinds of dung, and the species most typical of sheep dung. Ryndevich & Lundyshev (2005) recorded it from the nest of a song thrush. Common across much of Britain and Ireland in all kinds of grazing land, largely absent from the English east coast. *C. melanocephalus* ranges to Unst in the Shetlands and to St. Kilda and much of the Outer Hebrides, and to St. Mary's in the Scillies, and is also found on Jersey. Irish records include Rathlin and Lambay Islands. Recorded in all months, with peaks in May/June and in August.

12. *Cercyon nigriceps* **(Marsham)** Plate 83

Length 1.4-2.0 mm. *C. nigriceps* should be recognisable under the microscope as one of the two species, both small, with distinct femoral lines reaching almost to the front margin of the metasternum. Teneral specimens of *C. pygmaeus* may be as pale as *C. nigriceps*, but can be separated by the shape of the pronotal sides. At high power ($\times$ 80) it is possible to see fine hairs on the elytral interstriae. A form with an elytral spot, *centrimaculatum* Sturm, has been reported as common in Fermanagh (Johnson & Halbert, 1902). The origin of this species is uncertain. *C. nigriceps* was originally described from England but is regarded as a cosmopolitan synanthropic species associated with farm buildings and found in dung, mainly cow and horse, compost and other decaying matter. It flies both during the day and later to light, and may be found on window ledges and similar places, also having been caught in yellow pan traps and in an autocatcher. It is recorded from England in South Devon, East Sussex, West Kent, Surrey, South Essex, Hertfordshire, Middlesex, Berkshire, Oxfordshire, Cambridgeshire, East and West Suffolk, East Norfolk, Warwickshire, and Leicestershire, from Wales in Anglesey, from Scotland at Rannoch in Mid Perthshire, and in Northern Ireland from Antrim, Down and Fermanagh. It has recently been recorded from the Isle of Man. Records are from April to December, with a peak in May.

13. *Cercyon obsoletus* (Gyllenhal) Plate 84

Length 3.3-4.2 mm. *C. obsoletus*, incorrectly known in the past as *C. lugubris*, is a large and robust species, the only *Cercyon* known in Britain and Ireland with a small oblique depression just behind the outer corner of each elytron (Fig. 280). This ridge, the humeral plica, can also be found, though less well developed, in a recently described species, *C. castaneipennis* Vorst, 2009. The latter is found in the Netherlands (Vorst, 2009) and one might expect it to be found in southern England. *C. obsoletus* is dark with the rear of the elytra paler, plus a small pale spot on the elytral shoulder, whereas *C. castaneipennis* has markings rather like those of *C. haemorrhoidalis* (Fig. 264). *C. obsoletus* and *C. castaneipennis* also have the middle part of the "lyre" simple (Fig. 294), whereas this is lobed in all other species. Skidmore (1991) gave the habitat of *C. obsoletus* as the dung of cattle, deer and horses, but not that of
294   sheep. Ryndevich & Lundyshev (2005) recorded it from the nest of a song thrush. Hansen (1987) noted occurrence on exposed ground or in woodland, but not in shade, and indicated compost heaps, rotting vegetables and mushrooms, and carrion as alternatives to dung, confirmed by Vorst's (2009) list of habitats. *C. obsoletus* has a lowland distribution, frequent across the southern half of England, also on the Northumbrian coast and in Roxburghshire. Apart from Guernsey, Herm, Jersey, Ayrshire, Canna and North Aberdeenshire, the rest of the reported distribution is around the Irish Sea from Cumberland, the Isle of Man, Antrim, Armagh, Dublin, Waterford, Glamorgan, Pembrokeshire, Caernarfon, and Cheshire. Recorded from March to October with a peak in May.

14. *Cercyon pygmaeus* (Illiger) Plate 85

Length 1.2-1.8 mm. Distinguishing this species from *C. nigriceps* is discussed under that species. The conspicuous femoral lines should set this species apart from the similarly coloured and sized *C. terminatus*. This species occurs in all kinds of dung and in rotting plant material such as compost heaps. Ryndevich & Lundyshev (2005) recorded it from the nest of a fieldfare. Scattered across Ireland and Britain north to the Monach Isles, Canna, Raasay, East Inverness-shire and North Aberdeenshire, frequent only in areas subject to specialist recording. Other island records are for Anglesey, Bryher in the Scillies, the Isle of Man, Jersey, and Tiree. Recorded in all months except March, peaking in May and August.

15. *Cercyon quisquilius* (Linnaeus) Plate 86

Length 2.0-2.6 mm. *C. quisquilius* closely resembles the unspotted form of *C. unipunctatus*, from which it can be differentiated by the strength of the 9th and 10th elytral striae as in the key, and also by the mesosternal processes, narrow in both species but blunt at the rear in *C. unipunctatus* (Fig. 295) and sharply pointed in *C. quisquilius* (Fig. 296). The habitat is almost any decaying matter, but mainly fresh cow and horse dung. *C. quisquilius* is widely distributed in Britain, from South Hampshire to North Aberdeenshire, with a few clusters of records in areas of specialist recording. Welsh records are for Anglesey and Caernarfon, Irish records also being northern, from Antrim, East Donegal, Dublin, Fermanagh, and Louth. There are also records for the Isle of Man and for Jersey.
295   296   Recorded in all months except February, with a strong peak in August.

16. *Cercyon sternalis* (Sharp) Plate 87

Length 1.6-2.0 mm. The broad mesosternal process touching the metasternal process is typical of this species. The aedeagophore is similar to that of *C. convexiusculus* but with the median lobe longer, and to that of *C. sternalis* but with the extension to the basal piece broader. The secondary punctures on the elytra are exceedingly fine. *C. sternalis* occurs in a wide range of lowland freshwater habitats, also sometimes in brackish water: it is often found in association with tussocks and their associated litter. The English distribution is mainly in the south, north to Cheshire and Nottinghamshire, absent west of the Somerset Levels except on Bryher, St. Mary's and Tresco in the Scillies. In Wales *C. sternalis* is known from Anglesey and Monmouthshire, in Scotland from Kirkcudbrightshire in 2005, and on the Isle of Man from 1995. In Ireland this species was not known until 1996 but is now recorded from Clare, West Cork, North-east and South-east Galway, North and South Kerry, Louth, West Meath, Offaly, North Tipperary, and Wexford, all in the centre or in the south. *C. sternalis* was first reported on Jersey in 1998, on Alderney in 2011, and on Guernsey in 2013. Recorded in all months except January, peaking in May.

17. *Cercyon terminatus* (Marsham) Plate 88

Length 1.6-2.3 mm. The elytral pattern of this small species resembles that of *C. melanocephalus* but with the dark triangle typically larger and more diffusely edged, and with the suture pale even into the triangle. Found in association with all kinds of decaying matter, mainly fresh cow and horse dung. It has also been found on a dead seal on a Norfolk beach. Records for this species are scattered and mostly old. The English records are from St. Mary's in the Scillies, East Sussex, East Kent, Surrey, South Essex, Hertfordshire, Berkshire, Cambridgeshire, East Norfolk, Worcestershire, Warwickshire, Leicestershire, and South Northumberland. In Wales *C. terminatus* is known from Anglesey and Caernarfon and in Scotland from Dumfriesshire, Ayrshire, Lanarkshire, Selkirkshire and Dunbartonshire in the south and Raasay in the north. Irish records are largely coastal from Antrim, Armagh, East Donegal, Dublin and Sligo. There are old records from the Isle of Man and Jersey. Recorded in all months except January, with a peak in July.

18. *Cercyon tristis* (Illiger) Plate 89

Length 1.7-2.3 mm. The truncate parameres set this species apart from most other *Cercyon*, external characters being sufficient if used with care, particularly the primary elytral punctures, fading towards the rear unlike in the other three species with reticulate elytra. The non-reticulate *C. bifenestratus* also has truncate parameres but its basal piece is longer than the parameres. The species is found in a great range of manmade and natural freshwater habitats, sometimes a little brackish, and is mainly among wet plant debris. Grazing fen ditches are particularly favoured. Ryndevich & Lundyshev (2005) recorded it from nests of black-headed gulls and Buczyński & Tończyk (2010) note it from the nest of a common coot. This is mainly an eastern species in England, found in all lowland fen areas. The western distribution is patchier, frequent on the Cheshire and Solway Plains, and known west to South Devon, also on the Isles of Man and Wight, and Jersey. In Wales the distribution is also patchy, common on Anglesey, and recorded from Caernarfon, Flint, Monmouthshire, and Pembrokeshire. The Scottish distribution is even sparser, frequent in the mosses of Berwickshire and Roxburghshire, otherwise mainly coastal around the Solway, in Angus, Fife, East Lothian and Midlothian, East Inverness-shire and Ross, on Arran, the Monach Isles and South Uist. *C. tristis* ranges across Ireland from West Donegal to East Cork and Waterford, but it is frequent only in the centre. Recorded in all months except January, peaking in May.

19. *Cercyon unipunctatus* (**Linnaeus**) Plate 90

Length 2.4-3.4 mm. This is usually the most distinctively marked *Cercyon*. However, variants range from the dark var. *janssoni* Nyholm, with the black markings nearly as extensive as in *C. marinus* (Fig. 243) to a pale form var. *impunctatus* Kuwert (Lindberg, 1955): teneral specimens also cause confusion. It is regarded as a synanthropic species associated with rotting organic matter on farms, horse dung being specifically mentioned. Ryndevich & Lundyshev (2005) recorded it from nests of black-headed gulls, mute swans and a song thrush. This species can fly, in the evening at light, or during the day as has been reported in early spring in Scandinavia. *C. unipunctatus* has been most frequently recorded in England from Hertfordshire, Leicestershire, Nottinghamshire and Derbyshire, scattered outside this area from St. Mary's in the Scillies to East Kent in the south and to Bute, Eigg and Rum in the north-west and North Aberdeenshire and East Inverness-shire in the north-east, with outlying records from the Orkneys. Welsh records are for Anglesey and Glamorgan. Irish records are mainly for the north-east in Down, Londonderry and Louth, with an outlier in Limerick. There are records for the Isle of Man, Guernsey and Jersey. Recorded from February to November, with a peak in July.

Subgenus *Dicyrtocercyon* Ganglbauer

20. *Cercyon ustulatus* (**Preyssler**) Plate 91

Length 2.6-3.4 mm. The subgenus *Dicyrtocercyon*, of which this is the only species in Britain and Ireland, is characterised by the lateral profile being discontinuous, with the pronotum distinctly domed, sometimes as weak as in Fig. 228. The mesosternal process is narrow and narrowly separated from the metasternal process, which is distinctly pointed (Fig. 230). *C. ustulatus* is shining black, with a reddish yellow apical spot on the elytra split in two by the black sutural striae. The parameres are slightly flared at their tips, otherwise the aedeagophore has no distinguishing features. This species shares with *C. alpinus* and *C. impressus* a small groove on the pronotum just in front of the scutellum. *C. ustulatus* is mainly associated with muddy banks of streams and ponds amongst litter, and it can also be found beside water in cow dung. There is a Canadian record from a beaver lodge (Šmetana, 1979) and Ryndevich & Lundyshev (2007) record it in Belarus using the lodges of muskrats to hibernate and to pupate. Ryndevich (2008) adds the burrow of a common vole. *C. ustulatus* is common across lowland Wales and England, apart from the south-west beyond the River Teign, and occurs in southern Scotland north to Fife and Islay, other islands occupied being the Holy Island off of Arran and Little Cumbrae. *C. ustulatus* is also frequent in Ireland, with records from Antrim, Armagh, Cavan, Clare, West Cork, East and West Donegal, Down, Dublin, Fermanagh, South-east and West Galway, South Kerry, Leitrim, Laois, Louth, West Mayo, Monaghan, North and South Tipperary, Tyrone, Waterford, West Meath, and Wexford. Other islands with records are Anglesey, Jersey, Man and Rathlin. Recorded throughout the year, with peaks in May and September.

Subgenus *Paracercyon* Seidlitz

21. *Cercyon analis* (**Paykull**) Plate 92

Length 1.7-2.8 mm. The shape of this small species should be distinctive enough in the field, coupled with the strong link between the mesosternal process and the metasternal process. At high magnification (× 40 or more with diffuse illumination) it should be possible to see a tracery of fissures running mainly diagonally between the secondary

297

punctures of the elytral striae and issuing from the edge of the suture (Fig. 297): these could be misconstrued as the mesh-like microreticulation of *Cercyon* such as *convexiusculus* and *tristis*, but are more likely to go unnoticed. The habitat is in almost any rotting organic matter, as such sometimes caught in water, rarely in horse dung. It has been found in a sewage farm. Ryndevich & Lundyshev (2005) recorded it from nests of hooded crow, tufted duck, black-headed gull and mute swan, and they also (Ryndevich & Lundyshev, 2007) found it in lodges of beavers and muskrats. Records are scattered widely across lowland Britain from St. Mary's in the Scillies to Anglesey, Raasay, South Uist, Orkney and Shetland. Records are similarly scattered across Ireland from Antrim, East Donegal, Down, Fermanagh, South-east Galway, Londonderry, Roscommon, and south to Wicklow. There are records for the Isle of Man and Jersey. Recorded in all months except January and November, with peaks in May and September/October.

Subgenus *Paracycreon* d'Orchymont

22. *Cercyon laminatus* Sharp Plate 93

Length 3.2-4.0 mm. This large species is very distinct, with the black head and underside contrasting with the pale yellowish brown pronotum and elytra, often with the suture paler still. The head appears to be differently shaped to that of other *Cercyon*, because the eyes are larger and slightly protruding. The thin mesosternal keel and thin tips to the parameres are diagnostic. *C. laminatus* could be mistaken for a large teneral *C. littoralis*, which often has the head black and the body pale. Most specimens of *C. laminatus* are caught at light at dusk, the habitat appearing to be decaying organic matter, such as compost and bird faeces. Ryndevich (2008) also notes the nests of mute swan. The original record, from Japan, was in dung, and cow dung is specified as a habitat in Sweden. *C. laminatus* was first taken in England in 1959 in West Kent (Allen, 1969). It is now also known from South and North Hampshire, Surrey, Hertfordshire, Buckinghamshire, West Norfolk, Cambridgeshire, Bedfordshire, Northamptonshire, and Warwickshire. Recorded in May and from July to September with a strong peak in August.

16. *MEGASTERNUM* Mulsant

Megasternum are small hairless beetles, convex both dorsally and ventrally. They share with *Cryptopleurum* the mesosternum having a flat and wide pentagonally shaped process engaging in a notch on the rear edge of a hexagonal prosternal process. They differ from *Cryptopleurum* in having the outer face of the front tibiae strongly excavated. The dorsal surface has a largely effaced microreticulation that might cause confusion with the microreticulate *Cercyon* if the major differences in the undersides go unchecked. European *Megasternum* were long regarded as monospecific under the name *obscurum* (Marsham) and earlier as *boletophagum* (Marsham).

Males cannot be distinguished from females using external features. The long aedeagophore is best extracted by securing the specimen upside down and inserting fine forceps into the rear of the abdomen. The tips of the parameres are very thin and become distorted if stored dry: mounting in a drop of DMHF or a similar transparent mountant is advised.

That two forms are involved has only recently been recognised (Peter Hammond, pers. comm.; Welch, 2010; Lompe, 2012). Differences in the aedeagophore and the colour of the upper side of the body and the antennae appear to be the only characters of value. Other characters proposed to separate the forms have been found wanting:- the shape of the scutellum, more elongate in *concinnum*; the extent of the outermost row of punctures on the elytra, reaching more than halfway in *concinnum,* but found to be just as long in some *immaculatum*; the shininess of the elytra, i.e. the extent to which the microreticulation is effaced, found to be too variable; the extent to which the elytra taper to the rear, more so in *concinnum* but of no use without access to a range of specimens of both forms; the lengths of the metatarsus, found to overlap considerably. More work needs to be done to ratify separate species status, in particular analysis of DNA sequence data. The names employed here need to be formalised. They follow Welch (2010), based on the advice of Peter Hammond.

Key 12. The *Megasternum* complex

1. Body usually reddish-brown except for the darker head; antennal club usually as pale as the body; parameres slightly less than the length of the median lobe, which is narrowly constricted (Fig. 298); elytra tapering to the rear; length 1.7-2.2 mm .... ***Megasternum concinnum* (Marsham)**

- Body mainly black; antennal club typically black; parameres typically overarching the tip of the median lobe (Fig. 299), which is not as narrowed at the tip as in *concinnum*; elytra not tapering; length 1.8-2.0 mm .... ***2. Megasternum immaculatum* (Stephens)**

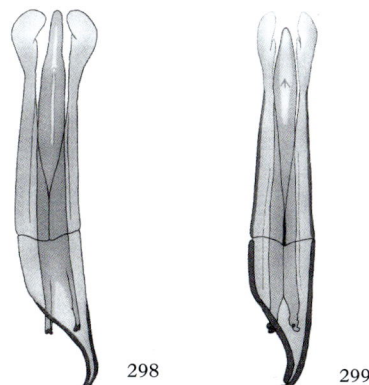

298   299

1. *Megasternum concinnum* (Marsham) Plate 94

Length 1.7-2.2 mm. Widespread in Ireland and Britain, extending at least to the Outer Hebrides and West Sutherland, also known from Jersey. Frequent among decaying vegetable matter and animal residues, such as are provided by grass tussocks, flood refuse, haystacks, and sheep dung, also reported from a mole's nest. The *Megasternum* larva was originally described (Phillips, 1923) from soil samples in permanent pasture in Ireland, being found in May, pupating in May and with adults emerging in June. Adults probably occur throughout the year, thus far noted from February to September.

2. *Megasternum immaculatum* (Stephens) Plate 95

Length 1.8-2.0 mm. Rarer than *concinnum* but widespread in Britain, probably also the same in Ireland, extending at least to the Outer Hebrides and West Sutherland. In the same habitats as *concinnum*, sometimes together with it: specified habitats include a dead squirrel, under fallen timber and in cow dung. Adults probably occur throughout the year, thus far noted from April to September.

17. *CRYPTOPLEURUM* Mulsant

This genus is distinguished from the similarly sized *Megasternum* by several characters, the most obvious being that the upper surface is finely hairy and that the front tibiae lack a notch. The elytra have 10 well incised and punctured striae, the interstices being heavily punctured.

External features should be distinct enough to avoid the need for dissection, but the aedeagophores can provide confirmation of an identification. The aedeagophore is like that of *Cercyon* and *Megasternum*, with a long and strongly curved stem to the basal piece. The parameres are narrow in two species and expanded and truncated in the other one. The last (fifth) visible abdominal sternite of *C. subtile* differs between the sexes, that of the female having a distinct tubercle in the middle.

Key 13. The species of *Cryptopleurum*

1. Body mainly reddish yellow with head and rear edge (sometimes also the middle) of the pronotum darkened; pronotum with an unusual form of sculpture, fine irregular longitudinal grooves running between the punctures, distinguishable from a network of microreticulation at × 60 or more (Fig. 300 shows detail); median lobe drawn out into a point and parameres less than the length of the lobe, their tips partly transparent, making it difficult to see that they are truncate (Fig. 301); length 1.4-2.2 mm ....
.... *Cryptopleurum subtile* **Sharp** (p. 94)

- Body mainly black with rear of elytra, and often a mark at the shoulder, dark red; pronotal microsculpture absent or, if present, a largely effaced microreticulate network; tip of median lobe more evenly pointed and parameres narrow (Figs 302 and 303) 2

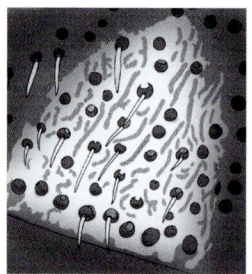

300

301

2. Elytral interstices raised but not developed into ridges; punctures of the interstices almost as strong as those of the pronotum; paramere tips very narrow (Fig. 302); length 1.3-2.2 mm .... **2. *Cryptopleurum minutum* (Fab.)** (below)

\- Interstices so strongly raised that those at the sides are almost keeled; punctures on the interstices finer than those of the pronotum; paramere tips slightly expanded and median lobe broader than that of *C. minutum* (Fig. 303); length 2.1-2.4 mm .... ***Cryptopleurum crenatum* (Panzer)** (below)

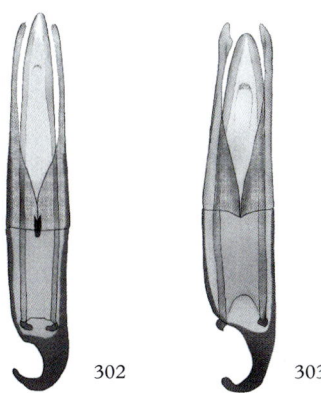

302   303

1. *Cryptopleurum crenatum* (Panzer) Plate 96

Length 2.1-2.4 mm. This species is most easily distinguished by its size and by the elytral interstices being raised, particularly so towards the sides. Assuming that *C. minutum* is available for comparison it appears easier to see these differences at <u>low</u> power (× 10). The aedeagophore has parameres wider at the tip than in *C. minutum*. A transverse channel running on the front of the head is almost entire in this species, whereas it is interrupted in *C. minutum*. *C. crenatum* can scarcely be described as synanthropic, not being entirely associated with decaying organic matter and dung, but also found in natural wetland habitats among plant debris and moss, particularly in sunlit sites according to Hansen (1987). When first brought forward as a British species, Tottenham (1939) noted records of *C. crenatum* from: North Wiltshire, Dorset, North Hants, West and East Sussex, Surrey, Berkshire, Cambridgeshire, Huntingdonshire, East Gloucestershire, Herefordshire, Worcestershire, Warwickshire, South-east, South-west and Mid-west Yorkshire, North Northumberland, and East Inverness-shire. Additional records are from the Scillies, South South Essex, Hertfordshire, Middlesex, and Oxfordshire, also from Wales in Caerfyrddin. Recorded in January, from April to June and from August to November, most frequent in May.

2. *Cryptopleurum minutum* (Fab.) Plate 97

Length 1.3-2.2 mm. This is an easily recognised and common species. *C. minutum* can be found in almost any rotting material. It is widely distributed across Ireland and Britain, known north to Raasay, Rum, East Inverness-shire and Moray. *C. minutum* is well recorded on islands, e.g. Holy Island in the Clyde Isles, Rathlin, the Isle of Man, Anglesey, the Scillies and Jersey. Recorded throughout the year, with a strong peak in May.

3. *Cryptopleurum subtile* **Sharp** Plate 98

Length 1.4-2.2 mm. *C. subtile* is distinctively coloured, red overall with at least the head and the rear pronotal edge darkened: the antennal clubs may be dark brown. The worm-like impressions running between the pronotal punctures are distinctive. This species, originally described from Japan, was first found in England, in South Devon in 1958 (Allen, 1984), and formally brought forward as British by Johnson (1967), who found it in Cheshire and Meirionydd. Additional records are from West Kent, Surrey, South Essex, Middlesex, Berkshire, East Norfolk, Northamptonshire, Worcestershire, Warwickshire, Cumberland, and the Isle of Man. There are also records for Scotland from Roxburghshire and Antrim in Ireland. Recorded in April, May and from July to January, with a peak in September.

18. *SPHAERIDIUM* Fabricius

Sphaeridium are often detected by their activity, either taking to flight or diving into liquid dung to avoid capture, unlike similarly sized and sluggish *Aphodius* species. Within the Hydrophilidae the spiny tibiae are also distinctive. As with the other medium-sized sphaeridiines in *Coelostoma* and *Dactylosternum* the eyes are notched at the front but *Sphaeridium* have eight-segmented antennae (Fig. 304) as opposed to the nine-segmented ones of the other two genera. The scutellum is peculiarly elongate (Fig. 19). Most specimens are brightly coloured, with red or yellow spots, the amount of pale colouring along the sides being helpful in identification. *Sphaeridium* are mainly associated with horse and cow dung (Hansen, 1987; Skidmore, 1991), being predators of fly larvae both as adults and as larvae. They can also be found in carcases, compost and other debris. *Sphaeridium* species are often found together.

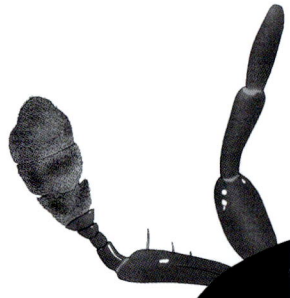

Figure 304. *Sphaeridium,* eight-segmented antennae.

Males are easily distinguished from females by their enlarged and strongly curved inner fore claws. Dissection of the aedeagophore is desirable to confirm the identity of *S. lunatum*, and it will also help with *S. bipustulatum* versus *S. marginatum*. The aedeagophore is dominated by the long median lobe, partly enfolded by the shorter parameres, and with a small basal piece.

Allemand and Leblanc (2004) followed the division of the genus into the subgenera *Sphaeridium* Fabricius for *lunatum* and *scarabaeoides*, and *Sphaeridionilus* Minozzi, 1921 for the other three species but these subgenera do not appear to be in use otherwise.

Key 14. The species of *Sphaeridium*

1. Hind angle of pronotum about 90° (Fig. 305); median lobe with a weak constriction slightly beyond the annulus and with a small projection at the tip (Figs 308 and 309); elytra typically with paler markings obscure, sometimes confined to the rear part of the suture and a small area around it 2

- Hind angle of pronotum much greater than 90° (Fig. 306); median lobe with a smooth outline (Figs 314 and 316), there being a very small projection at its extremity in one species (Fig. 314); elytra typically with bright red and yellow markings, and with the suture partly dark and dividing the apical spot in two 3

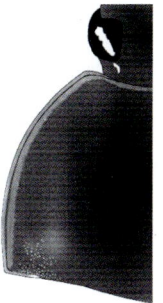
305

306

2. Elytra with rows of punctures on their sides, if not on the rest of the surface (Fig. 307 shows only the sites of the punctures of the rows, not the rest of the punctures); surface of elytra smooth or with weakly developed microreticulation; male inner fore claws small (Fig. 310) compared to those of *marginatum* (Fig. 311), their maximum dimension being just over half of the maximum width of the basal piece of the aedeagophore; sutural grooves continued around apex to outer rims of elytra (Fig. 307); aedeagophore (Fig. 308) smaller than that of *marginatum* (Fig. 309); length 4.2-6.5 mm ....
.... **1. *Sphaeridium bipustulatum* Fabricius** (p. 96)

- Elytra without punctures in rows and with much of the surface strongly microreticulate; male inner fore claws relatively large (Fig. 311), their maximum dimension about 0.7 times the maximum width of the basal piece of the aedeagophore; sutural grooves stopping near to apex in females (Fig. 312) but not in males; length 4.3-6.8 mm ....
.... **3. *Sphaeridium marginatum* Fabricius** (p. 97)

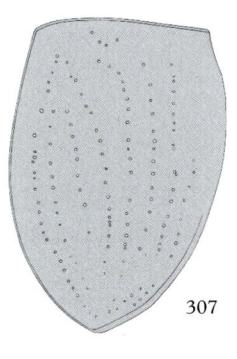
307

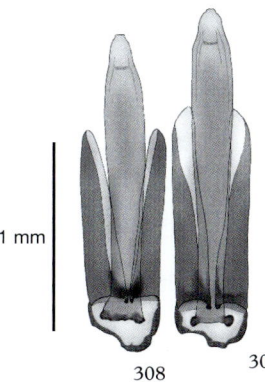
1 mm
308   309

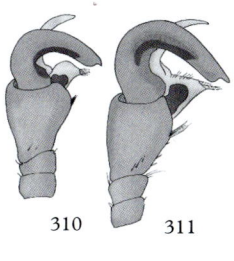
310   311

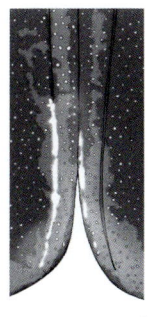
312

3. Pronotum entirely black; front margins of elytra and scutellum straight (Fig. 313); mid and hind femora black; aedeagophore with rounded end to the median lobe surmounted by a small process (Fig. 314); length 5.5-7.7 mm ....
.... **2. *Sphaeridium lunatum* Fabricius** (below)

- Pronotum with part of the margins yellow; shoulders of elytra with shoulders further forward than scutellum (Fig. 315); mid and hind femora reddish yellow or yellow each with a dark spot; aedeagophore with a sharp tip to the median lobe (Fig. 316); length 4.9-7.0 mm .... **4. *Sphaeridium scarabaeoides* (Linnaeus)** (p. 97)

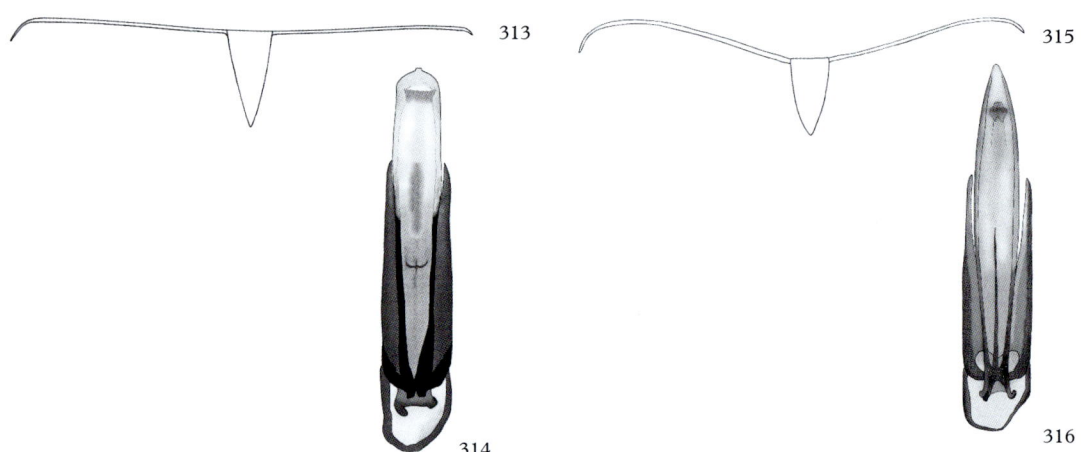

1. *Sphaeridium bipustulatum* Fabricius Plate 99

Length 4.2-6.5 mm. Despite its name, *S. bipustulatum* may be only obscurely marked with a reddish spot on each elytron. The yellow mark at the rear is not divided by a dark sutural stripe, but colouring is not a reliable guide for identification, and there are in any case structural features as described in the key. Old records of *S. bipustulatum* may be confused with *S. marginatum*, a species only reinstated as such comparatively recently (van Berge Henegouwen, 1989). *S. bipustulatum* appears to be more frequent than *S. marginatum* and is mainly English, recorded from North Devon, Isle of Wight, South Hampshire, West Sussex, East Kent, Surrey, North Essex, Berkshire, West and East Norfolk, Cambridgeshire, Herefordshire, Shropshire, and County Durham. There are confirmed records in Wales from Glamorgan, and in Scotland from Mid-Perth, but none as yet from Ireland. It is also recorded from the Isle of Man and from Alderney, Guernsey and Jersey. Recorded from April to September and in November, with peaks in April and August.

2. *Sphaeridium lunatum* Fabricius Plate 100

Length 5.5–7.7 mm. The entirely black pronotum, mid and hind femora set this species apart from *S. scarabaeoides*, as does the straighter edge to the front of the elytra. The shape of the tip of the median lobe is characteristic. For females the uneven bulging of the elytra provides a confirmatory character, the elytral edge being difficult to see from above except at the shoulder and at the rear, whereas the bulging is more even in *scarabaeoides*, making the elytral edge more generally visible. *S. lunatum* is frequent across Wales and England, less so on the coast, and commoner than *S. scarabaeoides* in the English Midlands. It reaches Easter Ross and North Aberdeenshire in Scotland, and is known from Canna, Islay and Raasay. Irish records, all modern, are from Antrim, Armagh, Cavan, Clare, West

Cork, County Down, South and North Kerry, Kilkenny, South Tipperary, Londonderry, and Longford. It is also known from the Isle of Man and Jersey. Recorded in December and January, and from February to October, with peaks in May and August.

3. *Sphaeridium marginatum* **Fabricius** Plate 101

Length 4.3-6.8 mm. The markings on this species may be reduced to a short section of the rear of the suture and a small area around it. Dissection of males is advisable, but care is needed to distinguish its median lobe from that of *S. bipustulatum*. This species appears to range more widely than *S. bipustulatum*, with which it was once confused (van Berge Henegouwen, 1989). *S. marginatum* was reinstated as British by Kirk-Spriggs and van Berge Henegouwen (1989) and is apparently more frequent than *S. bipustulatum* in the west. There are old records in Ireland from Antrim, Armagh, County Down, Dublin, and Laoise, and more modern ones from Fermanagh and South-east Galway. English records are from Dorset, South Hampshire, East and West Kent, Surrey, Middlesex, Berkshire, West Suffolk, Cambridgeshire, Huntingdonshire, Herefordshire, South Lincolnshire, Nottinghamshire, Cheshire, South-east Yorkshire, South Lancashire, County Durham, Westmorland, and Cumberland, and Welsh ones from Glamorgan and Caernarfon. Scottish records are for Ayrshire, Lanarkshire, Mid-Perthshire and Kincardineshire. Records are from April to September.

4. *Sphaeridium scarabaeoides* **(Linnaeus)** Plate 102

Length 4.9-7.7 mm. The margins of the pronotum are always to some extent yellow, and the yellow mid and hind femora each with a central black spot, provide a useful field character. *S. scarabaeoides* is the commonest *Sphaeridium* in Britain and Ireland. There are records in Ireland from Antrim, Cavan, Clare, East Cork, East Donegal, County Down, Fermanagh, South and North Kerry, and Londonderry. It is scattered across Britain, found particularly in coastal areas, notably common on Anglesey, in the Severn Valley and on the Cheshire Plain, and reaching north to West Sutherland, with records from Barra, Canna, Coll, Colonsay, Mull, Raasay, Rum, Skye, and Tiree in the Hebrides, also from Arran. *S. scarabaeoides* is known from the Isle of Man, and from Alderney, Guernsey, Jersey and Sark. Recorded in all months, peaking in May.

References and further reading

Allemand, R. & Leblanc, P. (2004). Identification des *Sphaeridium* de France (Coleoptera Hydrophilidae*). L'Entomologiste* **60**: 125-131.

Allen, A.A. (1969). *Cercyon laminatus* (Col. Hydrophilidae) new to Britain; with corrections to our list of species, and further notes. *Entomologist's Record & Journal of Variation* **81**: 211-216.

Allen, A.A. (1984). The earliest known British capture of *Cryptopleurum subtile* Sharp (Col.: Hydrophilidae). *Entomologist's Record & Journal of Variation* **96**: 35.

Allen, A.J. (2004). *Dactylosternum abdominale* (Fabricius) (Hydrophilidae) in Dorset – new to Britain. *The Coleopterist* **13**: 1-3.

Angus, R.B. (1977). A re-evaluation of the taxonomy and distribution of some European species of *Hydrochus* Leach (Col., Hydrophilidae). *Entomologist's Monthly Magazine* **112**: 177-201, 1 plate.

Angus, R.B. (1992a). Insecta: Coleoptera: Hydrophilidae: Helophorinae. *Susswaßerfauna von Mitteleuropa* **20** (10) part 2. Stuttgart: Gustav Fischer Verlag.

Angus, R.B. (1992b). A chromosomal investigation of *Helophorus brevipalpis* Bedel (Coleoptera: Hydrophilidae), with triploid Spanish females a possible source of American parthenogenetic material. *The Entomologist* **111**: 56-60.

Arribas, P., Velasco, J., Abellán, P., Sànchez-Fernández, D., Andújar, C., Calosi, P., Millán, A., Ribera, I. & Bilton, D.T. (2012). Dispersal ability rather than ecological tolerance drives differences in range size between lentic and lotic water beetles (Coleoptera: Hydrophilidae). *Journal of Biogeography* **39**: 984-994.

Balfour-Browne, W.A.F. (1958). *British Water Beetles*. **3**, Ray Society, London.

Balfour-Browne, F. (1962). *Water beetles and other things*. Dumfries: Blacklock Farries.

van Berge Henegouwen, A. (1986) Revision of the European species of *Anacaena* Thomson (Coleoptera: Hydrophilidae). *Entomologica scandinavica* **17**: 393-407.

van Berge Henegouwen, A. (1989). *Sphaeridium marginatum* reinstated as a species distinct from *S. bipustulatum* (Coleoptera: Hydrophilidae). *Entomologische berichten, Amsterdam* **49**:168-170.

Beutel, R.G. & Leschen, R.A.B. (eds) 2005. *Handbook of Zoology. Coleoptera, Beetles. Volume 1: Morphology and Systematics (Archostemata, Adephaga, Myxophaga, Polyphaga partim)*. Handbook of Zoology, Volume IV, Arthropoda: Insecta Part 38. Berlin: Walter de Gruyter.

Bilton, D.T. (1988). A survey of aquatic Coleoptera in Central Ireland and the Burren. *Bulletin of the Irish biogeographical Society* **11**: 77-94.

Bouchard, P., Bousquet, Y., Davies, A.E., Alonso-Zarazaga, M.A., Lawrence, J.F., Lyal, C.H.C., Newton, A.F., Reid, C.A.M., Schmitt, M., lipi ski, S.A. & Smith, A.B.T. (2011). Family-group names in Coleoptera (Insecta). *Zookeys* **88**: 1–972, doi: 10.3897/zookeys.88.807

Bratton, J.H. (1998). *Cercyon alpinus* Vogt (Hydrophilidae) in England. *Latissimus* **10**: 31.

Buczyński, P. & Tończyk, G. (2010). O kilku rzadkich i chronionych chrząszczach wodnych (Coleoptera) stwierdzonych w gniazdach ptaków i rybnych na Wyżynie Lubelskiej. *Wiadomo ci entomologiczne* **29**: 208-209.

Butler, P.M. & Popham, E.J. (1958). The effects of the floods of 1953 on the aquatic insect fauna of Spurn (Yorkshire). *Proceedings of the Royal Entomological Society of London* A **33**: 149-158.

Cox, M.L. (2007). *Atlas of the seed and leaf beetles of Britain and Ireland*. Pisces Publications.

Dobson, R.M., Hutchins, D. & Hancock, E.G. (2012). Notes on the occurrence and distribution of Coleoptera in Scotland associated with Rev. William Little. *Transactions of the Dumfriesshire and Galloway Natural History and Antiquarian Society* **86**:17-35.

Dudley, S.P., Gee, M., Kehoe, C., Melling, T.M., & the British Ornithologists' Union, Records Committee (2006). The British List: a checklist of birds or Britain (7th edition). *Ibis* **148**: 526-563.

Duff, A.G. (2013). 2012 Annual Exhibition Imperial College, London SW7 – 2 November. Coleoptera. *British Journal of Entomology & Natural History* **26**:42-43.

Duff, A.G., Campbell-Palmer, R. & Needham, R. (2013). The beaver beetle *Platypsyllus castoris* Ritsema (Leiodidae: Platypsyllinae) apparently established on reintroduced beavers in Scotland, new to Britain. *The Coleopterist* **22**: 9-19.

Fikáček, M., Prokin, M., Angus, R.B., Ponomarenko, A., Yue, Y., Ren, D. & Prokop, J. (2012). Phylogeny and the fossil record of the Helophoridae reveal Jurassic origin of extant hydrophiloid lineages (Coleoptera: Polyphaga). *Systematic Entomology* **37**: 420-447.

Foster, G.N. (2005). An annotated checklist of British and Irish water beetles, and associated taxa: Polyphaga, with an update to Adephaga. *The Coleopterist* **14**: 7-19.

Foster, G.N. (2010). *A review of the scarce and threatened Coleoptera of Great Britain. Part 3: water beetles.* Species Status No. 1. Joint Nature Conservation Committee, Peterborough.

Foster, G.N. & Friday, L.E. (2011). *Keys to adults of the water beetles of Britain and Ireland (Part I). (Coleoptera: Hydradephaga: Gyrinidae, Haliplidae, Paelobiidae, Noteridae and Dytiscidae).* Handbooks for the Identification of British Insects Vol. 4 Part 3 (2nd Ed.). St Albans: Royal Entomological Society. Preston Montford: Field Studies Council.

Foster, G.N. & Phillips, E.J. (1983). *Laccobius simulator* d'Orchymont (Coleoptera: Hydrophilidae) confirmed as British. *Entomologist's Gazette* **34**: 265-266.

Gentili, E. (1977). The British species of *Laccobius. Balfour-Browne Club Newsletter* **4**: 8-10, 13.

Gentili, E. (1988). Verso una revisione del genere *Laccobius* (Coleoptera, Hydrophilidae). *Osservatorio di Fisica terrestre e Museo Antonio Stoppani del Seminario Arcivescovile di Milano,* **9°** Annuario (n.s.) 1986: 31-47.

Gentili, E. (2006). Types de *Laccobius* Erichson, 1837 à l'Institut royal des Sciences naturelles de Belgique (Coleoptera, Hydrophilidae). *Bulletin de la Société royale belge de entomologie* **142**: 173-197.

Greenwood, M.T. & Wood, P.J. (2003). Effects of seasonal variation in salinity on a population of *Enochrus bicolor* Fabricius 1792 (Coleoptera: Hydrophilidae) and implications for other beetles of conservation interest. *Aquatic Conservation: Marine and Freshwater Ecosystems* **13**: 21-34.

Haliday, A.H. (1855). Entomological remarks. *In:* Anon. Proceedings. Dublin University Zoological Association June 2, 1855. *Natural History Review* **2**: 116-124.

Halstead, D.G.H. (1963). *Coleoptera: Histeroidea.* Handbooks for the Identification of British Insects Vol. 4 Part 10. London: Royal Entomological Society.

Hansen, M. (1982). Revisional notes on some European *Helochares* Muls. (Coleoptera: Hydrophilidae). *Entomologica scandinavica* **13**:201-211.

Hansen, M. (1987). *The Hydrophiloidea (Coleoptera) of Fennoscandia and Denmark.* Fauna entomologica scandinavica **18**, Leiden & Copenhagen.

Hansen, M. (1991). The Hydrophiloid Beetles. Phylogeny, Classification and a Revision of the Genera (Coleoptera, Hydrophiloidea). *Biologiske Skrifter* **40**: 368 pp.

Hansen, M. (1999). *World Catalogue of Insects Volume 2. Hydrophiloidea (s. str.) (Coleoptera).* Stenstrup: Apollo Books.

Hodge, P.J. & Jones, R.A. (1995). *New British Beetles. Species not in Joy's practical handbook.* Reading: British Entomological & Natural History Society.

Hubble, D. (2012). *Keys to the adults of seed and leaf beetles of Britain and Ireland.* Telford: FSC Publications.

Huijbregts, J. (1982). De nederlandse soorten van het genus *Cercyon* Leach (Coleoptera: Hydrophilidae). *Zoologische Bijdragen* **28**: 127-173.

Jackson, D.J. (1973). The influence of flight capacity on the distribution of aquatic Coleoptera in Fife and Kinross-shire. *Entomologist's Gazette* **24**: 247-293.

Johnson, C. (1967). *Cryptopleurum subtile* Sharp (Col., Hydrophilidae): an expected addition to the British list. *Entomologist* **100**: 172-173,

Johnson, W.F. & Halbert, J.N. (1902). A list of the beetles of Ireland. *Proceedings of the Royal Irish Academy* iii, **6**: 535-827.

Jones, R.A. (2002). *Tecticolous invertebrates. A preliminary investigation of the invertebrate fauna on green roofs in urban London*. Peterborough: English Nature.

Kirk-Spriggs, A. & van Berge Henegouwen, A. (1991). *Sphaeridium marginatum* in Britain. *The Balfour-Browne Club Newsletter* **49**:7.

Kuwert, A. (1886). General-Uebersicht der Helophorinen Europas und der angrenzenden Gebiete. *Wiener Entomologische Zeitung* **8**: 221-228, 247-250, 281-285.

Leach, W.E. (1817). *The zoological miscellany; being descriptions of new or interesting animals*. London: R.P. Nodder.

Levey, B. (2005). Some British records of *Chaetarthria simillima* Vorst & Cuppen, 2003 and *C. seminulum* (Herbst) (Hydrophilidae), with notes on their differentiation. *The Coleopterist* **14**: 97-99.

Lindberg, H. (1955). *Cercyon janssoni* Nyholm, en kustmelanistik form av *C. unipunctatus* L. *Notulae entomologicae* **35**: 68-71.

Löbl, I. & Šmetana, A. (eds) (2004). *Catalogue of Palaearctic Coleoptera, Volume 2 Hydrophiloidea – Histeroidea - Staphylinoidea*. Stenstrup: Apollo Books.

Löbl, I. & Šmetana, A. (eds) (2011). *Catalogue of Palaearctic Coleoptera. Volume 7. Curculionoidea I.* Stenstrup: Apollo Books.

Lompe, A. (2012). Gattung: *Megasternum* Muls. http://www.coleo-net.de/coleo/tabellen/body_megasternum.htm accessed 25 September 2012.

Luff, M.L. (2007). *The Carabidae (ground beetles) of Britain and Ireland*. Handbooks for the Identification of British Insects Vol. 4 Part 2 (2nd Ed.). St Albans: Royal Entomological Society.

Mann, D.S. (2006). *Ptilodactyla exotica* Chapin 1927 (Coleoptera: Ptilodactylidae: Ptilodactylinae) established breeding under glass in Britain, with a brief discussion on the family Ptilodactylidae. *Entomologist's Monthly Magazine* **142**: 67-79.

Morris, M.G. (2002). *True weevils (Part I). Coleoptera: Curculionidae (Subfamilies Raymondionyminae to Smicronychinae)*. Handbooks for the Identification of British Insects Vol. 5 Part 17b. London: Royal Entomological Society.

Morris, M.G. (2008). *True weevils (Part II) (Coleoptera: Curculionidae, Ceutorhynchinae)*. Handbooks for the Identification of British Insects Vol. 5 Part 17c. St Albans: Royal Entomological Society.

Owen, J.A. (1994). On the identification of *Cercyon alpinus* Vogt (Col: Hydrophilidae) and on its occurrence in Scotland. *Entomologist's Record and Journal of Variation* **106**: 181-183.

Phillips, K.C.J. (1923). The larva of a hydrophilid beetle, *Megasternum boletophagum*. *Irish Naturalist* **32**: 109-112.

Ryndevich, S.K. (2004). Review of species of the genus *Cercyon* Leach, 1817 of Russia and adjacent regions. I. Subgenus *Cercyon* (s. str.) Leach, 1817. *Cercyon lateralis*-group (Coleoptera: Hydrophilidae). *Annales Universitatis Mariae Curie-Skłodowska Lublin – Polonia* **59**: 29-41.

Ryndevich, S.K. (2008). Review of species of the genus *Cercyon* Leach, 1817 of Russia and adjacent regions. IV. The subgenera *Paracycreon* Orchymont, 1942 and *Dicyrtocercyon* Ganglbauer, 1904 (Coleoptera: Hydrophilidae). *Zoosystematica Rossica* **17**: 89-97.

Ryndevich, S.K. & Lundyshev, D.S. (2005). Beetles in birds' nests (Coleoptera: Noteridae, Dytiscidae, Helophoridae, Hydrophilidae & Dryopidae). *Latissimus* **20**: 17-19.

Ryndevich, S.K. & Lundyshev, D.S. (2007). Beetles (Coleoptera: Noteridae, Dytiscidae & Hydrophilidae) in muskrat and beaver lodges. *Latissimus* **23**: 28-30.

Schödl, S. (1991). Revision der Gattung *Berosus* Leach 1.Teil: Die palaarktischen Arten der Untergattung *Enoplurus* (Coleoptera: Hydrophilidae). *Koleopterologische Rundschau* **61** 111-135.

Schödl, S. (1993). Revision der Gattung *Berosus* Leach 3. Teil: Die paläarktischen und orientalischen Arten der Untergattung *Berosus* s.str. (Coleoptera: Hydrophilidae). *Koleopterologische Rundschau* **63** 189-233.

Schödl, S. (1998). Taxonomic revision of *Enochrus* (Coleoptera: Hydrophilidae). I. The *E. bicolor* species complex. *Entomological Problems* **29**: 111-127.

Shaarawi, F.A. & Angus, R.B. (1991). A chromosomal investigation of five European species of *Anacaena* Thomson (Coleoptera: Hydrophilidae). *Entomologica scandinavica* **21**: 415-426.

Shaarawi, F.A.I. & Angus, R.B. (1992). Chromosomal analysis of some European species of the genus *Georissus* Latreille, *Spercheus* Illiger and *Hydrochus* Leach (Coleoptera: Hydrophilidae). *Koleopterologische Rundschau* **62**: 127-135.

Shatrovskiy, A.G. (1984). Revision of the genus *Laccobius* Er. of the Soviet Union (Coleoptera, Hydrophilidae) [in Russian]. *Éntomologicheskoe Obozrenie* **63**: 301-325.

Short, A.E.Z. & Fikáček, M. (2011). World catalogue of the Hydrophiloidea (Coleoptera): additions and corrections II (2006-2010). *Acta Entomologica Musei Nationalis Pragae* **51:** 83-122.

Short, A.E.Z. & Fikáček, M. (2013). Molecular phylogeny, evolution and classification of the Hydrophilidae (Coleoptera). *Systematic Entomology* **38**: 723-752.

Skidmore, P. (1991). *Insects of the British cow dung community.* Occasional Publication No. **21**. Shrewsbury: Field Studies Council.

Šmetana, A. (1978). Revision of the subfamily Sphaeridiinae of America north of Mexico (Coleoptera: Hydrophilidae). *Memoirs of the Entomological Society of Canada* **105**: 1-292.

Šmetana, A. (1979). Revision of the subfamily Sphaeridiinae of America north of Mexico (Coleoptera: Hydrophilidae). Supplementum 1. *Canadian Entomologist* **111**: 959-966.

Šmetana, A. (1988). Review of the family Hydrophilidae of Canada and Alaska (Coleoptera). *Memoirs of the Entomological Society of Canada* **142**, pp. 316.

Southwood, T.R.E. & Johnson, C.G. (1957). Some records of insect flight activity in May, 1954, with particular reference to the massed flights of Coleoptera and Heteroptera from concealing habitats. *Entomologist's Monthly Magazine* **93**: 121-126.

Tottenham, C.E. (1939). *Cryptopleurum minutum* Fab. and *C. crenatum* Panz. (Col., Palpicornia, Sphaeridiinae). *Entomologist's Monthly Magazine* **75**: 117-118.

Vogt, H. (1971). 1. Unterfamilie: Sphaeridiinae. pp. 127-140 in: H. Freude, K.H. Harde & G.A. Lohse (eds) *Die Käfer Mitteleuropas.* 3. Krefeld: Goecke & Evers.

Vorst, O. (2009). *Cercyon castaneipennis* sp. n., an overlooked species from Europe (Coleoptera: Hydrophilidae). *Zootaxa* **2054**: 59-68

Vorst, O. & Cuppen, J.G.M. (2003). A third Palaearctic species of *Chaetarthria* Stephens (Coleoptera: Hydrophilidae). *Koleopterologische Rundschau* **73**: 161-167.

Welch, R.C. (2006). *Dactylosternum abdominale* (Fabricius) (Hydrophilidae) in East Northamptonshire – a second British locality. *The Coleopterist* **15**: 20.

Welch, R.C. (2010). Coleoptera recorded from Huntingdonshire during 2009, including ten species new to vice county 31. *Report of the Huntingdonshire Fauna & Flora Society* **62** (2009): 29-35.

West, A. (1930). Coleoptera. Chapter 40 in: A. S. Jensen, W. Lundbeck, T. Mortensen and R. Spärk (eds) *The Zoology of the Faroes*, Volume II, Part I (Crustacea, Myriapoda, Insecta I). Copenhagen: A.F. Høst & Son.

Whitehead, P.F. (2005). Notable Coleoptera records 5. *Entomologist's Gazette* **56**: 251-260.

Distributions

Each distribution is for Scotland (S), England (E), Wales, the Isle of Man (M), Northern Ireland (NI), the Republic of Ireland (RI), and the Channel Isles (CI), with records from 1980 onwards (●), and before 1980 (○).

	S	E	W	M	NI	RI	CI
Suborder Polyphaga							
HELOPHORIDAE							
Helophorus aequalis Thomson	●	●	●	●	●	●	●
Helophorus alternans Gené	?	●	●		○		●
Helophorus arvernicus Mulsant	●	●	●		●	●	
Helophorus brevipalpis Bedel	●	●	●	●	●	●	●
Helophorus dorsalis (Marsham)		●	●				
Helophorus flavipes Fab.	●	●	●	●	●	●	●
Helophorus fulgidicollis Motschulsky	●	●	●		●	●	○
Helophorus grandis Illiger	●	●	●	●	●	●	●
Helophorus granularis (L.)	●	●	●	●	○	●	
Helophorus griseus Herbst	●	●	●		●		
Helophorus laticollis Thomson		●					
Helophorus longitarsis Wollaston		●	●				
Helophorus minutus Fab.	●	●	●	●		●	●
Helophorus nanus Sturm		●	●			●	●
Helophorus nubilus Fabricius	●	●	●	○	○	○	
Helophorus obscurus Mulsant	●	●	●	●	●	●	●
Helophorus porculus Bedel	○	●	●	○	○	○	
Helophorus rufipes (Bosc d'Antic)	●	●			?	?	●
Helophorus strigifrons Thomson	●	●			●		
Helophorus tuberculatus Gyllenhal	○	●					
GEORISSIDAE							
Georissus crenulatus (Rossi)	●	●	●		●		
HYDROCHIDAE							
Hydrochus angustatus Germar	○	●	●	●		○	●
Hydrochus brevis (Herbst)	●	●	●		●	●	
Hydrochus crenatus (Fab.)			●				
Hydrochus elongatus (Schaller)	?	●	○				
Hydrochus ignicollis Motschulsky	●	●				○	●
Hydrochus megaphallus van Berge Henegouwen			●				
Hydrochus nitidicollis Mulsant			●				

	S	E	W	M	NI	RI	CI
SPERCHEIDAE							
Spercheus emarginatus (Schaller)		○					
HYDROPHILIDAE							
Hydrophilinae							
Anacaena bipustulata (Marsham)	●	●					●
Anacaena globulus (Paykull)	●	●	●	●	●	●	●
Anacaena limbata (Fab.)	●	●	●	●	●	●	●
Anacaena lutescens (Stephens)	●	●	●	●	●	●	●
Paracymus aeneus (Germar)		●					
Paracymus scutellaris (Rosenhauer)	●	●	●	○	●	●	
Berosus affinis Brullé	●	●	○				
Berosus fulvus Kuwert	●						
Berosus luridus (L.)	●	●	○			○	
Berosus signaticollis (Charpentier)		●	●			●	●
Chaetarthria seminulum (Herbst)	●	●				○	●
Chaetarthria simillima Vorst & Cuppen	●	●	●	●			●
Cymbiodyta marginellus (Fab.)	●	●	●	●	●	●	●
Enochrus affinis (Thunberg)	●	●	●				●
Enochrus bicolor (Fab.)	●	●	●				○
Enochrus coarctatus (Gredler)	●	●	●		●	●	
Enochrus fuscipennis (Thomson)	●	●	●		●	●	
Enochrus halophilus (Bedel)	●	●				●	●
Enochrus melanocephalus (Olivier)	●	●	●		●	●	
Enochrus nigritus Sharp		●	●				●
Enochrus ochropterus (Marsham)	●	●	●		●	●	
Enochrus quadripunctatus (Herbst)	●	●	●				●
Enochrus testaceus (Fab.)	●	●	●		●	●	●
Helochares lividus (Forster)	●	●					
Helochares obscurus (Müller)	●						
Helochares punctatus Sharp	●	●	●			●	●
Hydrobius fuscipes (L.)	●	●	●	●	●	●	●
Limnoxenus niger (Zschach)		●	●				●

	S	E	W	M	NI	RI	CI
Hydrochara caraboides (L.)	●	●					
Hydrophilus piceus (L.)	●	●					○
Laccobius atratus (Rottenberg)	●	●	●		●	●	○
Laccobius bipunctatus (Fab.)	●	●	●	●	●	●	●
Laccobius colon (Stephens)	●	●	●	●	●	●	
Laccobius minutus (L.)	●	●	●	●	●	●	●
Laccobius simulatrix d'Orchymont	●						
Laccobius sinuatus Motschulsky	●	●					○
Laccobius striatulus (Fab.)	●	●	●	●	●		
Laccobius ytenensis Sharp	●	●	●	●	●	●	●
Sphaeridiinae							
Coelostoma orbiculare (Fab.)	●	●	●	●	●	●	●
Dactylosternum abdominale (Fab.)				●			
Cercyon analis (Paykull)	●	●	●	●	●	●	●
Cercyon alpinus Vogt	●	●					
Cercyon bifenestratus Küster				●			
Cercyon convexiusculus Stephens	●	●	●	●	●	●	●
Cercyon depressus Stephens	●	●	●	●	●	●	●
Cercyon granarius Erichson				●			
Cercyon haemorrhoidalis (Fab.)	●	●	●	●	●	●	●
Cercyon impressus (Sturm)	●	●	●	●	●	●	●
Cercyon laminatus Sharp				●			
Cercyon lateralis (Marsham)	●	●	●	●	●	●	●

	S	E	W	M	NI	RI	CI
Cercyon littoralis (Gyllenhal)	●	●	●	●	●	●	●
Cercyon marinus Thomson	●	●	●		●	●	●
Cercyon melanocephalus (L.)	●	●	●	●	●	●	●
Cercyon nigriceps (Marsham)	○	●	●	●	●		
Cercyon obsoletus (Gyllenhal)	●	●	●	○	●	○	○
Cercyon pygmaeus (Illiger)	●	●	●	●	●	●	●
Cercyon quisquilius (L.)	●	●	●	●	○	●	○
Cercyon sternalis Sharp	●	●	●	●		●	●
Cercyon terminatus (Marsham)	●	●	●	○	○	○	○
Cercyon tristis (Illiger)	●	●	●	●	●	●	●
Cercyon unipunctatus (L.)	●	●	●	○	●	○	●
Cercyon ustulatus (Preyssler)	●	●	●	●	●	●	●
Megasternum concinnum s. lat.	●	●	●	●	●	●	●
Megasternum concinnum (Marsham)	●	●					●
Megasternum immaculatum (Stephens)	●	●	●				
Cryptopleurum crenatum (Kugelann)	○	●	○				●
Cryptopleurum minutum (Fab.)	●	●	●	●	●	●	●
Cryptopleurum subtile Sharp	●	●	●	●	●		
Sphaeridium bipustulatum Fab.	●	●	●				●
Sphaeridium lunatum Fab.	●	●	●	●	●	●	●
Sphaeridium marginatum Fab.	○	●	○	○	●	●	
Sphaeridium scarabaeoides (L.)	●	●	●	○	●	●	●

Appendix. Guidance on water beetles for non-specialists

This material has been updated from that presented in Part 1.

Aquatic bugs (Hemiptera)

Before going further make sure that the specimen to be identified really is a beetle. Aquatic bugs such as water boatmen can be mistaken for water beetles as they too have the front pair of wings modified into wing-cases. The most obvious difference between them and beetles is that these wing-cases overlap whereas they meet in the midline (or suture) in beetles. The overlap produces a diagonal cross (Fig. 317). Bugs have "incomplete metamorphosis", i.e. the immature stages resemble small versions of the adult, with the wings absent in the first instars and then appearing as wing buds after a few moults, so a beetle-shaped animal without elytra is likely to be an immature bug.

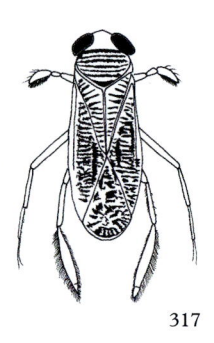

317

Ground beetles (Carabidae)

Ground beetles will fall into the water when shaken from emergent vegetation or splashed from the bank. With a few exceptions they stay on the water's surface and try to regain dry land. Some of the species associated with water (e.g. *Omophrum limbatum* Fabricius and *Oodes helopioides* (Fabricius) – see Luff (2007) for Plates 4 and 124 respectively) have an almost continuous body outline, but most have the pronotum and elytra of differing widths with the pronotum contracted at the rear (e.g. Fig. 318, *Agonum fuliginosum* (Panzer)). Ground beetles might be recognised by their running skill, even on water, but their ranges of colour and size (length 1.5-35 mm) can cause confusion. Other characters, not necessarily exclusive to them, are the thin, 11-segmented antennae, the robust mouthparts presented to the front, and the hind coxae modified into plates dividing the first abdominal segment. Some have a comb-fitted notch on the inner faces of the front tibiae (Fig. 319, the fore leg of *O. helopioides*) used to clean the antennae. A few species such as *Oodes helopioides* and *Carabus clatratus* Linnaeus deliberately enter the water. Carabidae are keyed by Luff (2007).

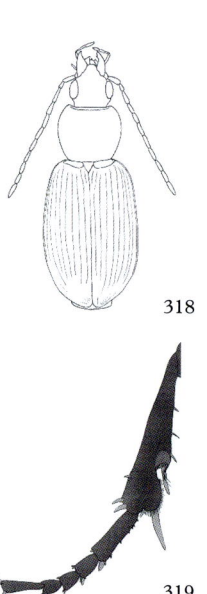

318

319

Key 15. The families of aquatic beetles

There is neither a single character nor a set of characters for recognising an aquatic beetle. Most beetles likely to occur in net samples are keyed below, which is intended mainly to help limnologists get to the right part of species identification keys. If a specimen will not key here, it has probably fallen in! It is advisable, once having decided the family to which a species belongs, to confirm this by calling up images of whole beetles on the web. This approach is most effective at the family and generic level, but can prove misleading for individual species identification.

A few ladybird beetles (Coccinellidae) are confined to wetland, in particular the *Coccidula* species and the nineteen-spot ladybird *Anisosticta novemdecimpunctata* (Linnaeus). The rove beetles (Staphylinidae) have many species associated with wet litter, and are beyond the scope of the current work. By contrast a largely aquatic family, the Ptilodactylidae, is represented by an introduced species confined to moist soil in greenhouses in England (Mann, 2006).

Users of this key are warned that the subject of this volume, the Hydrophiloidea, key out *on the very last line.*

1. At least four segments of the abdomen visible from above 2

- Not more than two segments protruding beyond elytra 3

2. Variously coloured elongate beetles, 1.5-20 mm long; antennae long and threadlike .... **STAPHYLINIDAE** rove beetles

- Orange-brown beetle, 2.2-3 mm long, louse-like (Fig. 320); antennae apparently in three parts with the last part made up of several fused segments – an insect living in the fur of beavers ....
  ... **LEIODIDAE or LEPTINIDAE** subfamily **PLATYPSYLLINAE**
  beaver beetle *Platypsyllus castoris* Ritsema (Part 3)

  First detected in the wild in Scotland in 2012 (Duff, 2013; Duff *et al.*, 2013).

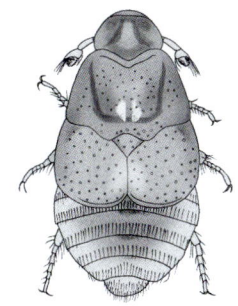

320

3. An elusive and exceptionally small beetle, 1.2 mm long at most, black and globular (Fig. 321), living on moist soil beside water .....
   .... **SPHAERIUSIDAE** (suborder MYXOPHAGA)
   microdot beetle *Sphaerius acaroides* Waltl (Part 3)

- All other beetles; if small, black and globular, then not less than 1.5 mm long .... Suborder ADEPHAGA and POLYPHAGA
  4

  There are other minute black beetles, some aquatic and some terrestrial, none of them being entirely globular with the elytra completely covering the abdomen.

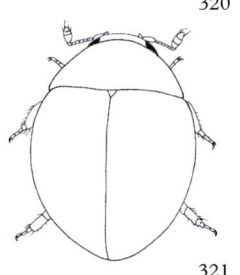

321

4. Middle and hind legs shorter than the front legs and very broad (Fig. 322); eyes divided horizontally to produce two pairs; can swim on the surface of the water; length 3.5-7.8 mm ....
   .... **GYRINIDAE** whirligig beetles (Part 1)

- Middle and hind legs as long as the front legs; eyes undivided; swim or float below the surface or walk on it; length 1-48 mm 5

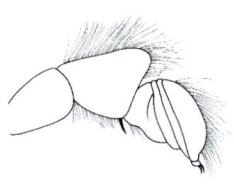

322

5. Antennae long and thread-like throughout most of their length, with 7-11 segments visible (Fig. 323), sometimes with the middle segments broader than the rest (Fig. 324) but never with a club at the end 6

- Antennae short with a terminal club (Fig. 325), or short with one or two basal segments much larger than the rest (Fig. 326) 13

If the antennae cannot be seen apply a fine brush to grooves on the front of the head in which they may be tucked

323

324

325

326

6. Antennae inserted on the front of the bead, the distance between their bases the same as or less than the length of their first segment (Fig. 327); most species found above the water, either more than 4 mm long with a metallic finish to the head, thorax and elytra or about 2 mm long and brown 7

- Antennae inserted further apart than the length of the first segment; mainly brown or black, with a metallic appearance being no more than a brassy finish to an otherwise black beetle; 1-38 mm long; species found above or below the water 8

327

7. Large, elongated beetles with the head prominent; elytra with at least five striae made up of rows of large punctures running between the centre and the shoulder, the additional stria next to the suture often being a short series of punctures on the front third of the elytron; species found above the water with metallic reflections; much rarer fully aquatic species with yellow and black or brown stripes; length 5-11 mm ....
.... **CHRYSOMELIDAE** Donaciinae or reed beetles
(consult Cox (2007) and Hubble (2012)

- Small, rounded beetle with head tucked in so far that it may invisible from above (Fig. 328); elytra with four striae as deep grooves between the centre and the shoulder on each elytron in addition to the sutural stria, which is usually confined to the front half of the elytron (Fig. 329); length 1.8-2.1 mm ....
.... **PSEPHENIDAE** water penny beetle
Eubria palustris Germar (Part 3)

328

329

8. Fourth segment of hind tarsus bilobed with the claw-bearing fifth segment rising from it (Fig. 330); elytra soft and well covered with hair; length 2.4-5.5 mm .... **SCIRTIDAE** marsh beetles (Part 3)

- Fourth segment of hind tarsus like the other non-claw bearing segments, not bilobed; elytra hard in mature beetles and with hair limited, usually on punctures in rows, or almost completely absent 9

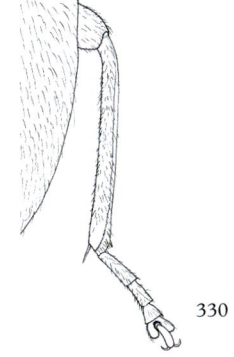

330

9. Pronotum as long as wide; head and eyes almost hidden below the convex front margin of the pronotum (Fig. 331) (beware alcohol-preserved specimens in which the head may protrude artificially); hind tarsi with the last segment very long and bulbous at tip (Fig. 332), not hairy; length 1.3-4.8 mm .... **ELMIDAE** riffle beetles (Part 3)

- Pronotum wider than long; head and eyes visible from above; last segment of hind tarsi neither very long nor bulbous, usually with a fringe of long swimming hairs; length 1.7-38 mm 10

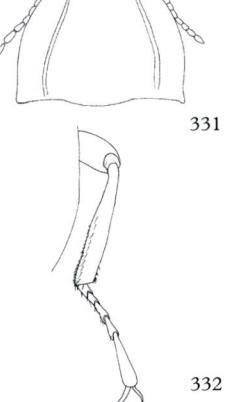

331

332

10. Hind coxae with large, rounded plates covering half of the abdomen and the basal half of the hind femora (Fig. 333); elytra each with about 10 longitudinal rows of large punctures, at least five of them on the dorsal side inside the shoulder; length 2.5-5.0 mm .... **HALIPLIDAE** crawling water beetles (Part 1)

- Hind coxae with lobed or pointed projections, most of the hind femora visible (Fig. 334); elytra each with not more than 5 longitudinal lines in total, usually made up of small punctures and/or grooves; length 1.7-38.0 mm 11

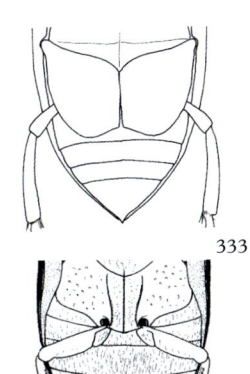

333

334

11. Head narrower than the front of the pronotum; body strongly convex below; elytra black with yellow or red front and side margins (Fig. 335); length 8.5-10.0 mm; noisily stridulates when alarmed .... **PAELOBIIDAE** squeak beetle
Hygrobia hermanni (Fab.) (Part 1)

- Head about the same width as the front of the pronotum and its outline usually forming a smooth curve with that of the pronotum (beware specimens preserved in alcohol where the head may be extended beyond its normal position); differing from the squeak beetle in the combination of colour and convexity 12

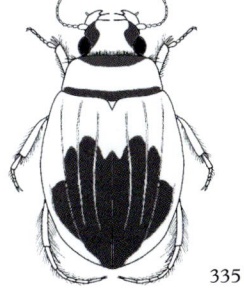

335

12. Hind coxal processes very broad (together at least as wide as long) forming a distinctive plate, the "noterid platform" (shaded in Fig. 336); middle antennal segments expanded (Fig. 337); length 3.5-5.0 mm .... **NOTERIDAE** burrowing diving beetles (Part 1)

336

337

- Hind coxal processes longer than broad, hind margin variously shaped (Fig. 338) but not as in the noterid platform; middle antennal segments usually longer than wide and thin; length 1.7-38 mm .... **DYTISCIDAE** diving beetles (Part 1)

338

13. Head with a snout or proboscis, at least as long as broad (examine from the side as well as from above); antennae long, with a definite "elbow" between the first and second segment in most species (Fig. 339) .... **CURCULIONIDAE** weevils – consult Morris (2002, 2008)

- No such snout, though the front of the head may bulge out in the centre; antennae without an "elbow" 14

339

14. Front and middle legs longer than the entire beetle (Fig. 340); antennae short .... **ELMIDAE** riffle beetles *Macronychus quadrituberculatus* Müller (Part 3)

- Front and middle legs shorter than the entire beetle; antennae long or short 15

340

15. Elytra and pronotum covered with dense hair; length 2.5-6.2 mm 16

- Elytra and pronotum without hairs, or with sparse hairs; length 1-48 mm 17

16. Antennae short, with an expanded second segment, often tucked into the front of the head (Fig. 341); last segment of the tarsi long; all tibiae slender; elytra uniformly greyish brown; length 3.8-5.4 mm .... **DRYOPIDAE** (Part 3)

341

- Antennae with first segment large, most of the rest forming a narrow club (Fig. 342); all tarsal segments short; front tibiae very broad; elytra patterned black or brown with yellow, coalescing flecks; length 2.5-6.2 mm .... **HETEROCERIDAE** (Part 3)

342

17. Palps from at least two-thirds to 4 times as long as the antennae and easily seen (Figs 343 and 344); antennae often hidden and each with a club with 3 or 5 enlarged hairy segments; length 1-48 mm .... "palpicorn" beetles
18

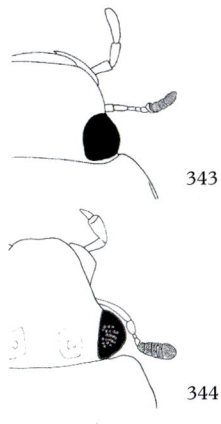

343

344

– Palps short, much less than two-thirds as long as the antennae, and not usually visible; antennae longer than the head and pronotum, with one segment in the club (Fig. 345); beetle convex below, black with yellow bases of the antennae and legs; length 1.5-1.8 mm .... **LIMNICHIDAE**
Limnichus pygmaeus (Sturm) (Part 3)

View from in front as well as from above.

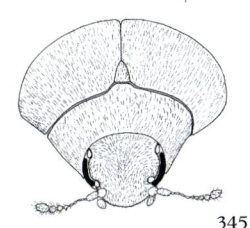

345

18. Small species, length 1.0-2.8 mm; antennae with the club having five finely hairy segments (Fig. 346); underside with 6 or 7 abdominal sternites visible .... **HYDRAENIDAE** (Part 3)

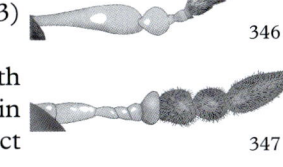

346

- Species varying in size from 1.3 to 48 mm long; antennae with the club having three enlarged and finely hairy segments as in Fig. 347 (or with a loose six-segmented club in one extinct species – see p. 40); underside with 5 visible sternites ....
.... **HYDROPHILOIDEA** (Part 2, p. 10 onwards)

347

Index

Main entries and start of main sections are shown in **bold**. Synonyms are given in *italics*.

Colour plates

Plate 1. Family Helophoridae
Helophorus nubilus Fabricius
2.8-4.0 mm (page 28)

Plate 2. Family Helophoridae
Helophorus porculus Bedel
4.0-5.0 mm (page 28)

Plate 3. Family Helophoridae
Helophorus rufipes (Bosc d'Antic)
4.5–5.5 mm (page 28)

Plate 4. Family Helophoridae
Helophorus tuberculatus Gyllenhal
2.8-3.8 mm (page 29)

Plate 5. Family Helophoridae
Helophorus alternans Gené
3.9-5.5 mm (page 30)

Plate 6. Family Helophoridae
Helophorus aequalis Thomson
4.5– 6.3 mm (page 30)

Plate 7. Family Helophoridae
Helophorus grandis Illiger
5.3-7.7 mm (page 31)

Plate 8. Family Helophoridae
Helophorus arvernicus Mulsant
2.4–3.5 mm (page 32)

Plate 9. Family Helophoridae
Helophorus brevipalpis Bedel
2.1-4.1 mm (page 33)

Plate 10. Family Helophoridae
Helophorus dorsalis (Marsham)
3.0-3.8 mm (page 31)

Plate 11. Family Helophoridae
Helophorus flavipes Fabricius
2.6-4.5 mm (page 31)

Plate 12. Family Helophoridae
Helophorus fulgidicollis Motschulsky
2.9-4.7 mm (page 31)

Plate 13. Family Helophoridae
Helophorus granularis (Linnaeus)
2.2-3.0 mm (page 32)

Plate 14. Family Helophoridae
Helophorus griseus Herbst
2.4-4.0 mm (page 32)

Plate 15. Family Helophoridae
Helophorus laticollis Thomson
2.9-4.2 mm (page 33)

Plate 16. Family Helophoridae
Helophorus longitarsis Wollaston
2.5-4.5 mm (page 33)

Plate 17. Family Helophoridae
Helophorus minutus Fabricius
2.4-3.4 mm (page 33)

Plate 18. Family Helophoridae
Helophorus nanus Sturm
2.2-3.0 mm (page 33)

Plate 19. Family Helophoridae
Helophorus obscurus Mulsant
2.4-4.3 mm (page 34)

Plate 20. Family Helophoridae
Helophorus strigifrons Thomson
3.2-4.2 mm (page 34)

Plate 21. Family Georissidae
Georissus crenulatus (Rossi)
1.4-2.1 mm (page 35)

Plate 22. Family Hydrochidae
Hydrochus angustatus Germar
3.0-4.0 mm (page 38)

Plate 23. Family Hydrochidae
Hydrochus brevis (Herbst)
2.8-3.7 mm (page 38)

Plate 24. Family Hydrochidae
Hydrochus crenatus (Fabricius)
2.1-3.1 mm (page 38)

Plate 25. Family Hydrochidae
Hydrochus elongatus (Schaller)
3.2-4.7 mm (page 39)

Plate 26. Family Hydrochidae
Hydrochus ignicollis Motschulsky
3.4-4.1 mm (page 39)

Plate 27. Family Hydrochidae
Hydrochus megaphallus van Berge Henegouwen
2.7-3.3 mm (page 39)

Plate 28. Family Hydrochidae
Hydrochus nitidicollis Mulsant
2.4-3.0 mm (page 39)

Plate 29. Family Spercheidae
Spercheus emarginatus (Schaller)
5.5-7.0 mm (page 40)

Plate 30. Family Hydrophilidae
Anacaena bipustulata (Marsham)
1.9-2.5 mm (page 43)

Plate 31. Family Hydrophilidae
Anacaena globulus (Paykull)
2.5-3.0 mm (page 43)

Plate 32. Family Hydrophilidae
Anacaena limbata (Fabricius)
2.4-3.2 mm (page 43)

Plate 33. Family Hydrophilidae
Anacaena lutescens (Stephens)
2.5-3.2 mm (page 44)

Plate 34. Family Hydrophilidae
Paracymus aeneus (Germar)
2.5-3.5 mm (page 45)

Plate 35. Family Hydrophilidae
Paracymus scutellaris (Rosenhauer)
2.5-3.2 mm (page 45)

Plate 36. Family Hydrophilidae
Berosus affinis Brullé
3.4-4.8 mm (page 48)

Plate 37. Family Hydrophilidae
Berosus luridus (Linnaeus)
4.1-4.8 mm (page 48)

Plate 38. Family Hydrophilidae
Berosus signaticollis (Charpentier)
4.8-6.1 mm (page 48)

Plate 39. Family Hydrophilidae
Berosus fulvus Kuwert
5.3-6.4 mm (page 49)

Plate 40. Family Hydrophilidae
Chaetarthria seminulum (Herbst)
1.3-2.0 mm (page 50)

Plate 41. Family Hydrophilidae
Chaetarthria simillima Vorst & Cuppen
1.3-3.9 mm (page 51)

Plate 42. Family Hydrophilidae
Cymbiodyta marginellus (Fabricius)
3.3=4.3 mm (page 51)

Plate 43. Family Hydrophilidae
Enochrus melanocephalus (Olivier)
4.2-5.1 mm (page 57)

Plate 44. Family Hydrophilidae
Enochrus bicolor (Fabricius)
6.5-7.5 mm (page 57)

Plate 45. Family Hydrophilidae
Enochrus fuscipennis (Thomson)
4.2-6.2 mm (page 58)

Plate 46. Family Hydrophilidae
Enochrus halophilus (Bedel)
4.8-5.9 mm (page 58)

Plate 47. Family Hydrophilidae
Enochrus ochropterus (Marsham)
4.7-5.5 mm (page 58)

Plate 48. Family Hydrophilidae
Enochrus quadripunctatus (Herbst)
4.7-5.7 mm (page 59)

Plate 49. Family Hydrophilidae
Enochrus testaceus (Fabricius)
5.4-6.7 mm (page 59)

Plate 50. Family Hydrophilidae
Enochrus affinis (Thunberg)
3.1-3.9 mm (page 60)

Plate 51. Family Hydrophilidae
Enochrus coarctatus (Gredler)
3.4-4.7 mm (page 60)

Plate 52. Family Hydrophilidae
Enochrus nigritus Sharp
3.2-3.9 mm (page 60)

Plate 53. Family Hydrophilidae
Helochares lividus (Forster)
4.5-5.8 mm (page 63)

Plate 54. Family Hydrophilidae
Helochares obscurus (Müller)
4.4-5.8 mm (page 63)

Plate 55. Family Hydrophilidae
Helochares punctatus Sharp
4.5-6.0 mm (page 63)

Plate 56. Family Hydrophilidae
Hydrobius fuscipes (Linnaeus)
6.4-8.7 mm (page 64)

Plate 57. Family Hydrophilidae
Limnoxenus niger (Gmelin)
8.0-9.8 mm (page 65)

Plate 58. Family Hydrophilidae
Hydrochara caraboides (Linnaeus)
13.0-18.5 mm (page 65)

Plate 59. Family Hydrophilidae
Hydrophilus piceus Linnaeus
34-48 mm (page 66)

Plate 60. Family Hydrophilidae
Laccobius atratus Rottenberg
3.1-3.8 mm (page 69)

Plate 61. Family Hydrophilidae
Laccobius bipunctatus (Fabricius)
3.2-3.8 mm (page 70)

Plate 62. Family Hydrophilidae
Laccobius simulatrix d'Orchymont
3-4 mm (page 70)

Plate 63. Family Hydrophilidae
Laccobius sinuatus Motschulsky
3.2-4.0 mm (page 71)

Plate 64. Family Hydrophilidae
Laccobius striatulus (Fabricius)
3.5-4.0 mm (page 71)

Plate 65. Family Hydrophilidae
Laccobius striatulus purpurascens Newbery
(page 71)

Plate 66. Family Hydrophilidae
Laccobius ytenensis Sharp
3.3-4.2 mm (page 71)

Plate 67. Family Hydrophilidae
Laccobius colon (Stephens)
2.4-3.1 mm (page 72)

Plate 68. Family Hydrophilidae
Laccobius minutus (Linnaeus)
2.6-3.2 mm (page 72)

Plate 69. Family Hydrophilidae
Coelostoma orbiculare (Fabricius)
4.0-4.8 mm (page 72)

Plate 70. Family Hydrophilidae
Dactylosternum abdominale (Fabricius)
3.8-5.0 mm (page 73)

Plate 71. Family Hydrophilidae
Cercyon alpinus Vogt
2.5-2.9 mm (page 83)

Plate 72. Family Hydrophilidae
Cercyon bifenestratus Küster
2.2-3.0 mm (page 83)

Plate 73. Family Hydrophilidae
Cercyon convexiusculus Stephens
1.6-2.2 mm (page 83)

Plate 74. Family Hydrophilidae
Cercyon depressus Stephens
1.8-2.6 mm (page 84)

Plate 75. Family Hydrophilidae
Cercyon granarius Erichson
1.7-2.4 mm (page 84)

Plate 76. Family Hydrophilidae
Cercyon haemorrhoidalis (Fabricius)
2.5-3.2 mm (page 84)

Plate 77. Family Hydrophilidae
Cercyon impressus (Sturm)
3.0-4.0 mm (page 85)

Plate 78. Family Hydrophilidae
Cercyon lateralis (Marsham)
2.5-3.2 mm (page 85)

Plate 79. Family Hydrophilidae
Cercyon littoralis (Gyllenhal)
2.5-3.3 mm (page 85)

Plate 80. Family Hydrophilidae
Cercyon littoralis a pale form
(page 85)

Plate 81. Family Hydrophilidae
Cercyon marinus Thomson
2.2-3.4 mm (page 86)

Plate 82. Family Hydrophilidae
Cercyon melanocephalus (Linnaeus)
2.3-3.0 mm (page 86)

Plate 83. Family Hydrophilidae
Cercyon nigriceps (Marsham)
1.4-2.0 mm (page 86)

Plate 84. Family Hydrophilidae
Cercyon obsoletus (Gyllenhal)
3.3-4.2 mm (page 87)

Plate 85. Family Hydrophilidae
Cercyon pygmaeus (Illiger)
1.2-1.8 mm (page 87)

Plate 86. Family Hydrophilidae
Cercyon quisquilius (Linnaeus)
2.0-2.6 mm (page 87)

Plate 87. Family Hydrophilidae
Cercyon sternalis (Sharp)
1.6-2.0 mm (page 88)

Plate 88. Family Hydrophilidae
Cercyon terminatus (Marsham)
1.6-2.3 mm (page 88)

Plate 89. Family Hydrophilidae
Cercyon tristis (Illiger)
1.7-2.3 mm (page 88)

Plate 90. Family Hydrophilidae
Cercyon unipunctatus (Linnaeus)
2.4-3.4 mm (page 89)

Plate 91. Family Hydrophilidae
Cercyon ustulatus (Preyssler)
2.6-3.4 mm (page 89)

Plate 92. Family Hydrophilidae
Cercyon analis (Paykull)
1.7-2.8 mm (page 89)

Plate 93. Family Hydrophilidae
Cercyon laminatus Sharp
3.2-4.0 mm (page 90)

Plate 94. Family Hydrophilidae
Megasternum concinnum (Marsham)
1.7-2.2 mm (page 91)

Plate 95. Family Hydrophilidae
Megasternum immaculatum (Stephens)
1.8-2.0 mm (page 92)

Plate 96. Family Hydrophilidae
Cryptopleurum crenatum (Panzer)
2.1-2.4 mm (page 93)

Plate 97. Family Hydrophilidae
Cryptopleurum minutum (Fabricius)
1.3-2.2 mm (page 93)

Plate 98. Family Hydrophilidae
Cryptopleurum subtile Sharp
1.4-2.2 mm (page 94)

Plate 99. Family Hydrophilidae
Sphaeridium bipustulatum Fabricius
4.2-6.5 mm (page 96)

Plate 100. Family Hydrophilidae
Sphaeridium lunatum Fabricius
5.5-7.7 mm (page 96)

Plate 101. Family Hydrophilidae
Sphaeridium marginatum Fabricius
4.3-6.7 mm (page 97)

Plate 102. Family Hydrophilidae
Sphaeridium scarabaeoides (Linnaeus)
4.9-7.7 mm (page 97)